Agricultural Waste Diversity and Sustainability Issues

Agricultural Waste Diversity and Sustainability Issues

Sub-Saharan Africa as a Case Study

Peter Onu

Department of Quality and Operations Management
Faculty of Engineering and the Built Environment
University of Johannesburg
Johannesburg, South Africa

Charles Mbohwa

Department of Quality and Operations Management
Faculty of Engineering and the Built Environment
University of Johannesburg
Johannesburg, South Africa

Academic Press is an imprint of Elsevier
125 London Wall, London EC2Y 5AS, United Kingdom
525 B Street, Suite 1650, San Diego, CA 92101, United States
50 Hampshire Street, 5th Floor, Cambridge, MA 02139, United States
The Boulevard, Langford Lane, Kidlington, Oxford OX5 1GB, United Kingdom

Library of Congress Cataloging-in-Publication Data
A catalog record for this book is available from the Library of Congress

British Library Cataloguing-in-Publication Data
A catalogue record for this book is available from the British Library

ISBN: 978-0-323-85402-3

For information on all Academic Press publications visit our website at https://www.elsevier.com/books-and-journals

Publisher: Megan Ball
Acquisitions Editor: Nancy Maragioglio
Editorial Project Manager: Cole Newman
Production Project Manager: Sruthi Satheesh
Cover Designer: Greg Harris

Typeset by TNQ Technologies

Contents

Preface

Diversity in agriculture is a concept recently gaining prominence. It requires improved approaches, which involve teamwork among numerous jurisdictions, departments within these jurisdictions, and collaborations between public and private entities to ensure sustainable development. It has been developed into a trendy word that is referred to whenever discussions focus on agriculture and food wastes or future perspectives in clean energy. Innovation provides better solutions that meet advanced, unaddressed, or existing market needs and hence the opportunity to consider a different way of thinking, consuming, or living. Up to the present time, the continual exploitation of fossil fuels as an energy source has led to numerous harmful effects on the environment, including a notable decrease in air quality (due to the greenhouse gases emitted by the combustion of hydrocarbons) and an increase in global temperature. Although there are already valid substitutes for fossil fuels from an energetic point of view, such as hydrogen, natural gas, and alcohol (such as methanol and ethanol), raw materials and economically convenient methods for producing these fuels in large quantities are lacking. As a consequence, this necessitates the need for dialogue about agricultural development, waste resource conservation, food industry operational effectiveness, and technological implementation to introduce innovations in biodiversity.

The aforementioned premise has been discussed not just as an opportunity but also as a precondition for assuring the sustainability in the agricultural sector. Moreover, scientists and engineers are trying to optimize production processes starting from lignocellulosic biomasses not coming from crops, biodegradable wastes from food manufacturing processes, and wastes from farming operations. Like most agricultural development literature, the text begins by discussing the scientific, engineering, and technological principles underlying waste streams. These discussions inform the reader of each waste stream's risks, with an eye toward precautionary, reliable, and resilient approaches to reduce these risks. The text culminates with recommendations and ideas about best practices and proper management of wastes before, during, and after they are generated. We give considerable attention to treatment and conversion, but best management practices involve the reduction and elimination of waste volume in its various forms, sectors, and streams.

This book aspires to fill in this gap by providing an understanding of the implementation of strategies in agricultural development activities for sustainable gains. This book is split into 3 different sections and 10 chapters. The first part is an introduction to wastes management in the agriculture and food sector, including methodological approaches in the preparation and processes of the waste involved. The second part is a collection of chapters on the most important energy generation techniques and strategies. The last section of this book covers best practices and management: sustainability, associated technologies, accountability, communications, and involvement on diverse stakeholders. It illustrated the use of a mathematical model to minimize operational costs in the agrowaste management process and discussed the concept of ecoefficiency. The section focusses on the prospect of agrowaste management and risk associated in the sub-Saharan African region with an emphasis on Nigeria, Uganda, and South Africa (in the West, East, and Southern part of the continent—Africa).

The authors share certain solutions to overcome limitations, guide developmental processes, and improve academia/entrepreneurial interactions and knowledge transfer in agricultural wastes management and development. In essence, no single resource can be sufficiently comprehensive on its own in dealing with waste, so each of the chapters is richly annotated and sourced to give the reader ample avenues for further investigation. Hence, the book is a contribution to the body of literature that supports waste management and innovative protocols for achieving sustainability. We strived to include environmental and engineering issues daunting emerging sustainability and technologies that contribute to the diversity and livelihood of the sub-Saharan African region. Importantly, the usefulness of this book calls for unequivocal attention to address the challenges of how to achieve effective and sustainable solutions to one of the most quintessential and dynamic regional problems (agricultural waste management).

The book is summarized as follows:

CHAPTER 1: WASTE MANAGEMENT AND THE PROSPECT OF BIODEGRADABLE WASTES FROM AGRICULTURAL PROCESSES. Agricultural waste diversity and sustainability issues have become a serious concern lately that have led to huge financial and environmental implications in developing countries and sub-Saharan Africa in perspective. The lack of proper wastes management practices, following the lack of adequate information, and compliance with established protocols has become a challenge too great to be downplayed. The chapter introduces waste management strategies to promote agricultural business development. The ideology of recovery, recycling, and reuse in waste management processes is discussed. It contributes to agricultural waste management operational strategies, with inferences to technologies and viable crops that can promote food and energy security.

CHAPTER 2: AGRICULTURAL WASTES AND OPPORTUNITIES FROM FOOD PRODUCTION CHAIN. The world population growth rate directly influences the increase in food productivity and agricultural expansion within and outside the boundaries of a nation, and Africa is significantly affected by this narrative. The need to assimilate the potential of the various waste management processes and optimization for value creation, replacing significant portions of fossil fuels by alternative energy and a variety of spinoffs, including eco-by-products useful for domestic and industrial purposes is timely. The chapter details waste streams and strategies that access operational effective practices in food production operations, which directly affects the output of resources coming from the farms.

CHAPTER 3: METHODOLOGICAL APPROACHES IN AGROWASTE PREPARATION AND PROCESSES. Sustainable innovations and strategies for natural resource utilization, without causing any significant damage to the ecological balance, promoting water, land, and forest conservation are vital subjects to be addressed. To protect the ecosystem, there is a great need to adjust our current lifestyle and explore sustainable resources for biofuel production. Emerging techniques and conservative practices that can promote the ethanol extraction process from cellulosic and biodegradable materials are presented in the chapter. The operational principles in detoxification, hydrolysis, and fermentation and the significance of biofuel production have been discussed.

CHAPTER 4: SUSTAINABLE AGRICULTURAL WASTE DIVERSITY: ADVANCES IN GREEN ENERGY AND MATERIAL PRODUCTION. Waste to energy is actualizable, especially with the new trend in research and development that has brought about a nascent improvement in the classification of energy crops. The concern for waste to energy has, therefore, become of vital interest to different private, public, and government stakeholders, leading to numerous funding programs primarily to boost the Africa stance on sustainable agroecology. The chapter explores the techniques and technologies of producing bioenergy (biodiesel), biomaterial from food, and agrowaste. The need for a sustainable approach to address these threats of resource depletion and environmentally unsafe practices in Africa cannot be downplayed. The chapter contends with sustainable industrial agriculture solutions for developing countries and sub-Saharan Africa in perspective.

CHAPTER 5: SUSTAINABLE AGROWASTE DIVERSITY VERSUS SUSTAINABLE DEVELOPMENT GOALS. The average amount of waste generated annually from agriculture and food processing amounts to 1.3Gt globally. This is an obvious challenge that requires a timely strategy to mitigate the negative impacts on natural resources conservation. The motivation of this research is to examine the relationship, between the agrowaste management system, with inference to agriculture development, and the prospect of achieving the United Nations' Sustainable Development Goal (SDG). We gather that SDG 7 stands out the most concerning sustainability and agricultural waste diversity, whereas SDGs 5, 11, and 14 least support the proposition.

CHAPTER 6: NEW APPROACH AND PROSPECTS OF AGROWASTE RESOURCES CONVERSION FOR ENERGY SYSTEMS PERFORMANCE AND DEVELOPMENT. The development of renewables and alternative technologies to ensure eco-friendly measures and meet the production of natural gas shows considerable growth potential for the electrification of modern cities and isolated or rural communities. Biofuels are achieved from oil-based producing crops and other biodegradable agricultural residues and biomass feedstocks that have undergone several processes to yield alcohol in the form of ethanol. The chapter addresses the occurrence, techniques, and strategies of biofuel production, with inference to the classification and characteristics of the fuels. A concentrated assessment of agrowaste conversion has been reported, offering vital contributions to the understanding of the properties of biofuels.

CHAPTER 7: OVERVIEW OF MODELS FOR AGRICULTURAL WASTE MANAGEMENT AND TRENDS IN BIOFUELS PRODUCTION. The need to pursue waste recovery is multifaceted, relating to clean energy generation, management of unwanted and discarded items, emission control, and maximization of the use of land against uncontrolled dumping. The motivation of the study is within the purview of agricultural development economics and touches on the lackadaisical coordination of developing countries regarding sustainable agrowaste management practices. The chapter is a contribution to the opportunities and challenges of energy production from biodegradable wastes. We demonstrate using mathematical modeling to show how operations cost can be minimized in a biofuel station in Uganda, as a case study.

CHAPTER 8: NASCENT TECHNOLOGIES IN RESOURCES CONSERVATION AND SUSTAINABLE AGRICULTURAL DEVELOPMENT. Agriculture diversity has become an important concept lately. It supports social, economic, and environmental improvement. It also ensures ecoefficiency within the agroindustrial sector and promotes sustainable values. The chapter focuses on ecoefficiency techniques, applying the stochastic frontier approach (SFA) for environmental improvement activities through the application of green technology. It covers a scenario of integrating clean energy, power generating systems to facilitate farm activities while ensuring accurate data management. In essence, the interlink between conservative management and nascent technology innovations as a strategy for social, economic, and environmental performance is presented. The contribution will support future researchers and guide decisions on green initiatives.

CHAPTER 9: WHICH WAY FORWARD: AGRICULTURAL WASTES MANAGEMENT AND THE FOURTH INDUSTRIAL REVOLUTION APPRAISAL. To improve sustainability through agrowaste diversity, the agricultural and food industry sectors need to reform their strategies and access to digitized operations management practices (ecoliteracy). Already, several notable agricultural development companies have embraced digitization and knowledge transfer capabilities that can promote operational effectiveness in biodegradable wastes management dispositions. The chapter discusses knowledge management concerns in conjunction with the Industry 4.0 promulgation for agricultural development solutions. The authors analyze emerging technologies and nascent initiatives in agrowaste management, with emphasis on sustainability and conservatism. The chapter recommends several technologies and sustainable initiatives to promote agricultural waste diversity.

CHAPTER 10: ECONOMICS AND RISK ASSESSMENT OF NEW TECHNOLOGIES IN AGROWASTE DIVERSITY. This timely research examines the interlink between sustainability and the ideals for economic progress amidst the risk and uncertainty that threatens the advancement of agriculture diversity in Africa. Inferences to decision-making parameters based on the availability of new technologies and the implementation of Sustainable Development Goals (SDGs) in the sub-Saharan African region from the viewpoint of promoting livelihood is presented. The chapter concludes with the corroboration of the SDG 8 and its impact on the achievement of the SDGs 1, 2, 3, 4, 5, 6, 7, and 15. The authors juxtapose sustainable interventions that should proffer competitive solutions to specific societal, social, economic, and environmental challenges. The study recommends several technologies and sustainable initiatives to promote agricultural waste diversity.

Acknowledgment

"I have not attempted to cite in the text all the authorities and sources consulted in the preparation of this work. To do so would require more spaces than is available. The list would include departments, libraries, industrial institutions, periodical, and many individuals." However, my gratitude goes to Kampala International University, Ahmadu Bello University, and staff and management of the University of Johannesburg for the research experience. I am thankful to my family and all well-meaning friends and well-wishers who continue to support me.

CHAPTER

Waste management and the prospect of biodegradable wastes from agricultural processes

1

1.1 Introduction

Agriculture wastes are solid or liquid materials generated during agricultural activities, from direct consumption of primary products or their industrialization, which is no longer useful for the process that generated them (Onu & Mbohwa, 2020c). The concept of waste management takes several dimensions; reduction, reuse, and recycling of the generated wastes (Kumar, Dhingra, & Singh, 2018; Skaggs, Coleman, Seiple, & Milbrandt, 2018). Wastes generated from homes, restaurants, schools, factories, and small or large businesses, to mention a few, result in hundreds of millions of tons each year (Liuzzi, Sanarica, & Stefanizzi, 2017). Academics and managers are keen to improve operations in the area of agriculture, especially in light of the growing global food crises and environmental degradation challenges threatening livelihood. The most important of which include globalization and intense competition for conservatism and sustainability, to allow positive changes in the ecosystem (Yusuf et al., 2019).

Waste management seeks to redress human responsibility to environmental safety and thus protects life and properties and promotes healthy living (Akiyode, Tumushabe, Hadijjah, & Onu, 2017). The presence of solids, liquids, and gaseous waste in high concentration caused by anthropogenic and industrial manufacturing operations leads to pollution and in the long run poses significant risks, with a multiplier effect that affects livelihood (Onu & Mbohwa, 2019a, 2019b, 2019c, 2019d; Sheahan & Barrett, 2017). Solid waste is defined as any waste that is dry or in a semisolid form and is disposed of as undesired residues. The majority of solid wastes are generated by households as refuse or as no-value domestic waste. Other categories of solid waste disposal come from industries, commercial institutions, and agricultural activities (Pradhan & Mbohwa, 2014; Yahya, 2018). These wastes include plants shaft/leaves/straws, rice, corn, and other plant husks shells; fruits peels, seeds, and sugarcane bagasses. Such wastes are predominant in sub-Saharan Africa, where there are paralytic waste management practices (Chifari, Lo Piano, Bukkens, & Giampietro, 2018; Woldeyohannes, Woldemichael, & Baheta, 2016).

Agricultural Waste Diversity and Sustainability Issues. https://doi.org/10.1016/B978-0-323-85402-3.00006-1

Agriculture and waste management

Generally, waste is classified based on their form of existence or occurrence. Organic wastes from agricultural activities, municipal, industrial, biomedical, and business enterprise waste all exist in solids, partial solid to the liquid state, whereas the nature of existence of these residues could be available in a biodegradable or nonbiodegradable form (as in solids), such as waste from human activities, farming operations, and food processing, which are discarded (Ngoc & Schnitzer, 2008; Nwakaire, Ezeoha, & Ugwuishiwu, 2013; Onu & Mbohwa, 2018a). Domestic wastes such as sullage, human and animal/poultry excrement, and industrial wastewater that are nontoxic undergo biodigestive processes, to convert them for gainful purposes, as by-products that become raw materials for green energy development (Engg, 2013; Prapinagsorn, Sittijunda, & Reungsang, 2017; Xu, Fu, Yang, Lu, & Guo, 2018). A popular definition, according to the OECD: "Agricultural waste is waste produced as a result of various agricultural or farming operations. It includes manure and other wastes from farms, poultry houses and slaughterhouses; harvest waste; fertilizer runoff from fields; pesticides that enter into the water, air or soils; and salt and silt drained from fields" (Jones & Schoonover, 2010; Liebetrau, Sträuber, Kretzschmar, Denysenko, & Nelles, 2019).

Consequently, they may pose nonhazardous threats (organic nature) to the environment and could also hold great economic potential (Bajwa, Peterson, Sharma, Shojaeiarani, & Bajwa, 2018; Saikia, 2011). However, the environmental impact of agricultural wastes ranges from the effects due to greenhouse gas emissions that occur at landfills and the consideration for apportioning of space to initiate biodiversity of the wastes (Yusuf et al., 2019; Manyuchi, Mbohwa, & Muzenda, 2017; Mooney et al., 2009). The outlook for agricultural expansion soon draws interest to effective management techniques, which consider waste elimination otherwise, which may lead to catastrophic challenges (Lawal et al., 2019; Scarlat, Motola, Dallemand, Monforti-Ferrario, & Mofor, 2015). These shortcomings include having to deal with smells or the adverse effect of greenhouse gas emission and toxic runoff that affects useful fertile lands and the water ecosystem (Gomiero, 2018; Pattanaik, Pattnaik, Saxena, & Naik, 2019; Pollet, Staffell, & Adamson, 2015).

Continuous development in technologies linked to the agriculture business in these modern days, such as the Internet of things (IoT), is playing an essential part in everyday life by expanding our insights and our ability to engage in precision farming. This can contribute to effective farming and lead to waste elimination. Technological advancement helps assess agricultural variables such as soil conditions, the biomass of animals or plants, and weather conditions. The ranking order for waste management while eliminating the disadvantageous effects that impede the sustainability pursuit and certainly add value to clean energy production while ensuring minimal disposal is seen in Fig. 1.1, thus prioritizing reduction and total avoidance of wastes generation.

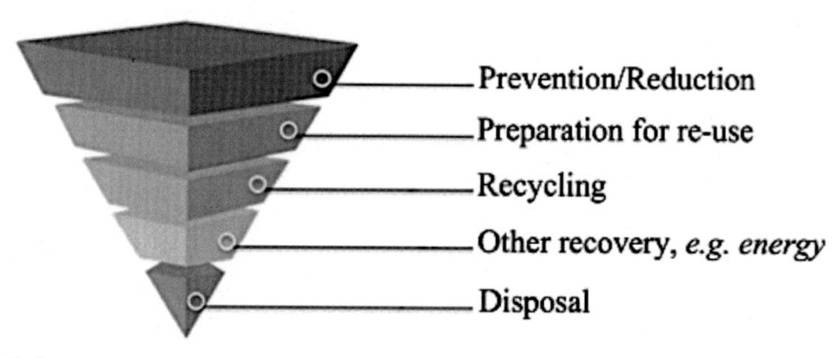

FIGURE 1.1

Waste management hierarchy.

Sustainable development and the role of agrowaste management

Sustainable development has become a popularly accepted concept that has gained wide perspectives over time. The United Nations commission first published a report on the subject matter, in the year 1987, which posits that sustainable development is: "development that meets the needs of the present generation without compromising the ability of future generations to meet their own needs" (Brundtland, 1987). The report had a focus on three central issues, listing economic growth, environmental conservation, and social equality. It, therefore, necessitates the urgency for changing culture toward waste management or the integration of developmental strategies with environmental conservation policies. This was targeted at converting agriculture wastes (Gutiérrez, Serrano, Siles, Chica, & Martín, 2017) into harmless by-products or exploiting their energy potential for economic and social performance (Onu & Mbohwa, 2018b). A few corporate social performances are discussed.

Reduce poverty

The global population index continues to rise, regardless of the existing mediocre social, economic, and environmental performance in developing economies, and calls for joint intervention to win the war against poverty (Akiyode et al., 2017; Lebreton & Andrady, 2019; Prakash, 2011). Africa needs to intensify her efforts as per waste collection, recycling, and reconversion, which is lucrative for income generations, and creates opportunity for employment. The practice of waste reduction may require expert human capability or machine endeavor as a conservative strategy for sustainable operation (Ikhlayel, 2018). Increasing agricultural productivity in a zero-waste generation operation or the reclamation of "good waste" creates the avenue where food products become cheaper and accessible and lessen the bar on poverty as it brings about a change in the standard of living for the consumers. In contrast, the profit margins also increase for the agriculturist (Ferranti, 2019). Agricultural waste control and effective food storage processes stand to expand the horizon to curb poverty. Moreover, the agricultural sector today drives many economies and surely guarantees more reliably jobs.

End hunger

The worldwide campaign against food wastes to end hunger has become acceptable and calls for policy implementation to support its course, all over the world. The agroindustry, especially, the food process and production company, generates enormous amounts of wastes in their different capacities, depending on the operation being carried out. The selection process of worthy of sale goods and the discarding of several food items as waste is discouraged nowadays (Compton, Willis, Rezaie, & Humes, 2018; Mao, Feng, Wang, & Ren, 2015). New regulations prohibit farmers from burning, burying, or disposing of agricultural waste incessantly, without consideration of their viability and potential to feed the underprivileged. The majority of the food waste formed from fruit peels and wastewater from food processing operation is used to produce syrup and flavoring, respectively, among other products. Development of a variety of food crops that can withstand the harshest conditions and survive preservation duration has become the order of the day, hence ensuring high, healthy yields that are affordable (Bond & Morrison-Saunders, 2011; United Nations (UN), 2015; Woldeyohannes et al., 2016).

Good health and well-being

The campaign to proffer balance to the different global environmental issues calls for proper waste disposal or management, which helps to limit the volume of harmful resources released to the environment (Ikumapayi, Akinlabi, Onu, Akinlabi, & Agarana, 2019; Onu, Abolarin, & Anafi, 2017). The measure can contribute to the reduction of carbon monoxide and carbon dioxide emission that emanates from fossil fuel activities through various bioremediation strategies and waste conversion technologies to generate energy (Woldeyohannes et al., 2016). Preservation of the earth's ecosystem is promoted in the absence of toxic wastes that litter the environs and damage fertile lands. The commercialization and expansion of waste management have the potential to promote healthy lifestyles and create new job opportunities from the design and marketing of enabling technologies, to maintenance operations carried out on the systems (Onu & Mbohwa, 2019e; Zabed, Sahu, Boyce, & Faruq, 2016).

Clean water and sanitation

The category of solid, liquid, and gaseous waste indiscriminately dumped and poorly managed becomes a potential threat and poses a significant risk to human and all forms of the living organism. Direct and indirect pollution occurs at different intervals of wastes collection, treatment, or disposal (Dong et al., 2016; Song, Wang, Li, & Duan, 2012). In most cases, the pollution is tracked from a single point or an abundant source and in other scenarios is formed by different events and unknown sources (fertilizer from farms, debris forest, agricultural food processing wastes, industrial sewage, etc.), which is usually transported by different agents and deposited into domestic water sources (lakes and streams), thus polluting the water bodies and making them unsafe to use while still destroying the habitat within (aquatic lives).

Affordable and equitable energy

The disposition to populate sustainable development actions, with the focus on alternative and renewable energy options, presents a reconversion of agricultural wastes for their energy potential. Industrial sewage, wastes from processing operation, and municipal (organic) and biomedical wastes are regarded as a viable source for sustainable clean energy. The idea is to stimulate the development of biorefineries to convert these categories of wastes and proffer nascent solutions to exploit energy from them (biodegradable and nonbiodegradable) (Leichenko, 2011; Rothaermel & Hess, 2010). Moreover, existing carbon credit deals between developing African nations and counterpart nations, on the financing of renewable energy and agricultural waste digesters/systems ought to be revised to capture recent events/development. The global impacts of greenhouse gas are carefully sorted through the exploitation of alternative clean energy for domestic and industrial usage, having considerable little or no effect on the climatic condition (Onu & Mbohwa, 2020d). Furthermore, access to bioconversion technologies domesticated to serve in residential capacity is proficient at reducing the number of municipal wastes which goes out, while the conversion plant meets some of the energy (heat) demands at homes and eateries (Achinas & Euverink, 2016; Jönsson, Alriksson, & Nilvebrant, 2013).

Industrialization and innovation

The different techniques adopted for arid farming operations and technologies used are fast changing. Improvement in agricultural extension and efficient operations for irrigation farming, chemical spraying consideration, and veterinary services, among others, has become more action focused on reducing the impact of agricultural waste. Different innovative approaches are as follows: educating and training, implementing new technology, and improving old infrastructure to feet new design requirement, and cultural change is directed toward gainful utilization and reduced risk of agricultural waste (International Energy Agency, 2016; United Nations (UN), 2015). Discoveries to access and detect early danger signs of biological wastes adopt a detection and measurement instrument "biosensor," which can track and inform the toxicity level and hazard characteristic of residues. Newly emerging technologies have capabilities that allow for optimal conversion of agrowastes through improved frameworks of collection, separation, digestion, and slurry treatment. Moreover, the application of advanced digital techniques to monitor operations and ensure zero wastes production in agriculture has inspired research (Meena et al., 2019; Song et al., 2012).

Sustainable production and consumption approach to agricultural waste management

The prevention of chemical or harmful pollutant released into the air, water, and soil, thus, promotes safe and healthy habitation of all living organism, thus leaving the lands fertile (free from erosion), the air free from dangerous gases and aquatic life forms sustain healthily (Raji & Onu, 2017). Small farmers play a vital role to apply adequate knowledge of conservative agricultural practices, supportive of renewable energy diversity to lead climatic change mitigation. The business of wealth creation from agricultural waste conversion has become a new area where both large and small enterprises venture, thus encouraging the application of sustainable initiatives to effectively collect, process, and convert agricultural wastes. More so, by-products of these processes have been used for different purposes (production of composite material and aquaculture feeds to mention a few). Managing farm wastes would no doubt ensure environmental protection and food security.

A forecast by the Food and Agriculture Organization (FAO) predicts an average of food production to rise to 40% by 2030 and 70% by 2050 to mitigate the growing population trend, double the value of current energy and water demand by 2050 (Food and Agriculture Organization of the United Nations, 2017). The implication is that increasing agriculture and food production may lead to more wastes being generated. It is vital to improve production and consumption style to ensure food security and social and environmental performance, hence ensuring human survival in the years to come by applying ecologically focused long-term approach to manage resource. Feeding practices need to change, and product design can be revolutionized to meet the eco-friendly biodegradable requirement. Less synthetic and more organic products, especially in the supply chain process, are advised for correctional remedies to reduce the impact of agricultural wastes.

1.2 Agricultural waste management operations strategy

The availability of wastes from farms, and unwanted discards from food production industries, or the selection of lignocelluloses-based crops that can be exploited for sustainable energy disposition, depends on the geoecology and techniques used to convert the wastes. This practice has been reviewed in Chapters 4 and 5. The biochemical and thermochemical pathways for value-added products from waste beneficiation operations have also been detailed in the said chapters.

The major challenges in agriculture in developing nations and rural areas are as follows:

- **i.** Quality of crops and land available for farming and agriculture.
- **ii.** Lack of availability of utilities such as electricity, machinery, and dependable irrigation systems.
- **iii.** The low economic condition of farmers, and lack of government support or incentives.
- **iv.** The disparity between product prices from farmers to customers. The involvement of a third party (retailer) has increased the gap between what farmers are paid and what customers pay for.
- **v.** The geographical and climatic conditions have an unpredicted adverse effect on the farming process and outcomes.

vi. Lack of technology in agriculture either due to lack of access, high cost of commissioning, and operation or due to knowledge.
vii. Lack of public enlightenment and the inadequate or proper medium to disseminate information concerning waste management potentials and benefits.
viii. Bias against waste management processes.

It is vital to access the key players in the decision-making process (cultural change, regulations, and project survival tendencies, etc.) for agricultural development. When the cost of identifying and financing the appropriate services or systems demand to meet sustainability expectations falls within reconcilable benefits, the confidence to implement agrowaste management becomes high. However, readily available and accurate information concerning waste repository and the network of the collection is quintessential for an exploitation strategy to convert waste to gain. It is reassuring to say that international interest toward sustainable development permits nascent research and funding to notable universities, globally to remediate agricultural wastes matters in general, with emphasis on agrowaste management and with particular interest for rural development. Guerrero, Maas, and Hogland (2013) identified generation and separation, collection, transfer and transport, treatment, disposal, and recycling as the essential factors, which influence waste management systems as shown in Fig. 1.2.

Generation and separation

The level of awareness of farm operators and the farm size (rural settlement or remote country farm area) are elements which directly and indirectly affect waste generation. Also, the decision to prepare and separate the different categories of wastes generated must be guided by the educational background and information on sustainability achievement from agricultural waste management. The nature of

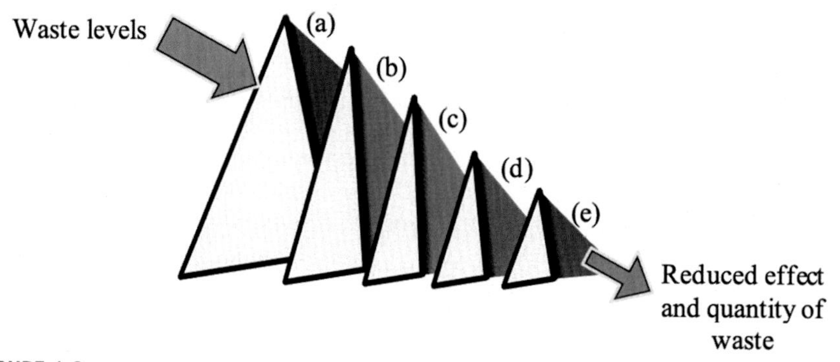

FIGURE 1.2

Factors influencing elements of waste management systems and expectations. (A) Waste generation and separation, (B) waste collection and transfer/transportation, (C) waste treatment, (D) waste disposal, (E) waste recycling.

farm and cultivation activities or products, and by-products, based on the competitiveness and technology application can affect the concentration and characteristics of waste being produced (Amsterdam & Thopil, 2017). However, most farmers manage their generated waste through new and emerging technology, using advanced automobiles, digital sensors, and other automated systems configuration to ensure optimal performance in waste elimination (Compton et al., 2018).

Collection, transfer and transport

The collection and movement of waste are greatly dependent on the technicality involved, the processes and mechanisms required to be put in place to successfully execute proper waste collection (Carrillo-Nieves et al., 2019; Eisted, Larsen, & Christensen, 2009). The financial determinant supports waste collection technologies and information systems to deal with waste management schedule: time (availability vs. demand), route, customer satisfaction, and operational equipment underpin on the planning strategy required to ensure effective agrowaste management. Separation, collection, and transportation of the different waste categories demand a well-organized protocol, amid sound routing plans. This is to ensure adequate information flow, which outlines the parameters that affect waste availability, generation, and expected target to meet agricultural waste endearment. Hence, the most convenient practice of waste control and the method for separation, based on the social, environmental, and economic viability, must be applied by rural farmers and isolated island dwellers that engage in commercial farming to translate the essence of resource control and sustainability advancement.

Treatment

The essential technologies, infrastructure, and guiding policies for handling and agricultural diversity, needed to remediate agrowaste challenges in rural settlements, are nonexistent in Africa (AGRA, 2017; Yin, Stecke, & Li, 2018), whereas in other cases, the discomfort caused during the process (smelly fumes—thermal) leads to infectious spreads. The sorting and preparing period of wastes for reuse precedes the biological waste treatment (composting or digestion process), where viable by-products are obtained. Different world projects and international funding, which focuses on Africa waste treatment through biodigesters in rural settlement to control widespread diseases, have suffered neglect. Poor communication and the lack of consistency in the coordination and management of wastes have failed in several projects. As such, training on technologies, technical strategies, and behavioral adaptation of stakeholders on agricultural development is essential to maintain a conservative ecosystem and contribute to sustainable development.

Disposal

The initiative to have waste accessors and recovery options for elimination purposes is rather emphasized. A zero-waste agricultural network that promotes contact between SMEs and farm operators to boycott burning (incineration), burying (land-filling), composting, and vermicomposting or in the case of disposal at sea is encouraged. Hence, innovative pathways to make useful the wastes from agricultural processes should continue to evolve and meet the changing times (O'Sullivan, Bonnett, McIntyre, Hochman, & Wasson, 2019; Seadon, 2010). Agricultural waste disposal tactics that involved the requisition of containers and bins over short distances between farms affects the costs of the general operation of waste management, together with the cost of transportation and handling the wastes (Compton et al., 2018). Thus, on-site miniplants, agrowaste resource conversion, and reconversion (waste generated from processed waste) within proximity to farms are gaining advantage toward product design, as in the case of raw materials or composite development, and wealth creation.

Recycling

The social challenges and the cost of waste disposal or the underutilization of green waste have sprung up the market for recycling with nascent techniques about the way to go about equipment and processing machinery. The "West" and the Europeans, for example, already have actualizable policies/legislations for waste recycling and funding for managing the operational activities. The former is a result of the sustainable potential from waste management from the perspective of socio-economic and environmental development. The focus point on the recycling of agro-waste looks to the nonbiodegradable types, i.e., hard exocarps, tree backs, branches, etc., and in most cases also, the nature of sludge, which is recycled to produce domestic nonharmful auxiliaries such as cardboard, additives for pottery (molding), and container to mention few (Dong et al., 2016). Financing support that will increase information dispersion and grooming professionals to become agrowaste recycle pioneers to reach the farthest and nearest waste potential centers and to exploit and maximize the benefits of these wastes is required (Andriani, Wresta, Saepudin, & Prawara, 2015; Bruce et al., 2018). Hence, there is need to apply low-cost recycling technologies in the short run, for developing economies, while nascent innovations and active collection protocols are desired to bridge the gap of distance, collection capacity per period, farmers' interest and participation, and infrastructure for housing the collected wastes.

The problem of seasonality significantly affects agrowastes exploitation and leads to operational redundancy of the technologies, which is used to create viability; several agricultural wastes are only available in specific seasons or a specific time of the year. This yardstick, coupled with the volume of the agricultural waste, or the rate of the garbage generation at any instance, when not sufficient, leaves waste conversion facility less effective. Flexible designs of robust bioplant converter are required to operate efficiently to meet the various challenges of agro-waste management. Also, the aforementioned is essential for a centralized location strategy to quickly and smartly collect wastes from different networks and hot spots.

1.3 Sustainable technology entry versus agricultural diversification

Cost-effective, high-performance techniques and technologies are what are needed to service municipalities and cities, especially the low-income and developing countries, as per agrowaste management. Exploiting the economic importance of agrowastes is beneficial and promotes profitability, societal and social satisfaction, and environmental protection (Dahiya et al., 2018). Coupled with the obvious, regarding urbanization and population increase, and the threat of increased consumption and production that conversely ascribe to high risk of waste generation, authorities must devise sustainable means to manage what is considered as waste. While emerging technologies and techniques are springing up, old methods of waste management are being reconsidered to ensure value creation (Onu & Mbohwa, 2019g). Nonetheless, as the idea of waste management becomes increasingly accepted, attention is given to agricultural cash and food produce as per consumption purposes or used in renewable energy production. Regardless, waste management must be treated with high regard as the contrary may lead to an adverse socioenvironmental hazard (Bing et al., 2016; Song, Fisher, & Kwoh, 2019).

Importantly, based on the key underlying factor for agricultural development outlook in the sub-Saharan Africa region and the economic pretext to increase the chances for production, domestic demand, and trade of agricultural commodities, the projection by the "FAO" shows a stagnated growth (2020–25) in per capita GDP over the next 5 years, resulting in a significant slowdown in the sub-Saharan African region as seen in Fig. 1.3. This in turn affects the decision for investment and optimal resources management. As such, the more likely reason to intensify agricultural resources diversity for sustainable deductions thus stimulates agricultural biomass research to meet sustainable and renewable energy and material

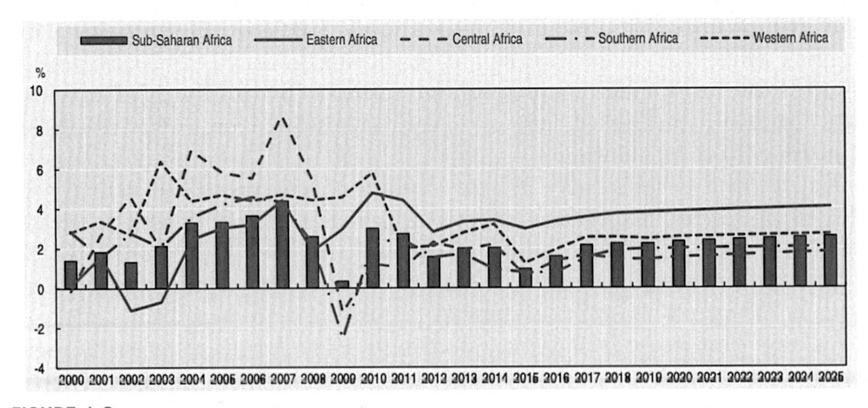

FIGURE 1.3

GDP growth per person in sub-Saharan Africa (Agriculture in Sub-Saharan Africa: Prospects and challenges for the next decade, 2016).

demand for the future (Shedrack et al., 2019). Accordingly; advances made in science that can transform agriculture business, making it profitable while assuring sustainability, have been discussed (Long, Blok, & Coninx, 2016; Pierpaoli, Carli, Pignatti, & Canavari, 2013). Technologies that reduce the burden on farmers in that sophisticated machinery now performs tedious jobs that ordinarily would have been carried out by farmers now exist for future agricultural development (Pivoto et al., 2018; Qu, Wang, Kang, & Liu, 2018).

New technology implementation and innovation in agrowaste management

The desired framework for sustainable technology intervention is required to influence the pace of implementation of policies for agrowaste conversion. Favorable regulatory requirements for the commercialization of suggested agricultural recourses and biomasses for energy production will significantly affect the success of the future biofuel market (Glasziou et al., 2014; Panesar, Kaur, & Panesar, 2015). It will substantially encourage competitiveness among green energy companies. However, the production of a viable biofuel from potentially significant agricultural resources means high development costs, which unfold new challenges. Bridging the gap in research between academia and green energy, industries aim to resolve these limitations and offer a unique opportunity for innovative biofuel production. The pursuit of technological solutions that combine low cost and excellent operational tendency must inspire new methods, products, and services (Onu & Mbohwa, 2018b, 2019b, 2019c; Onu & Mbohwa, 2019f, 2019h). This will transform the agribusiness sector and raise the urgency to strategic ways to reduce wastes and maximize energy savings and cost, which will require wholistic technological applications (Tepic, Fortuin, Kemp, & Omta, 2014). Suitable techniques and sustainable technologies must be utilized throughout the farming operation, from production through processing, distribution, and retailing, up to the end of consumption stages.

To ensure effective waste management processes that will lead to sustainable value creation, a high degree of care is required from the start of the waste beneficiation process, which involves the use of several technologies, some of which are applied in the preliminary stages or pretreatment operations. These operations include hydrothermal treatment, ozonation, microwave, ultrasound, and enzymatic hydrolytic pretreatment (Lynd et al., 2017; Zabed et al., 2016). However, not all of the aforementioned technologies are fully developed or fall within the confinement of preference, as per the type of technology that matches the requirement for selection from a geographic perspective. The idea of waste management and the waste to wealth contribution, which suffice through research, have become the order of the day and have created potable technology designs of decentralized units, which promotes the network for energy producers (Meneses-Jácome et al., 2016; Onu & Mbohwa, 2020b, 2020a).

A brief on biodegradable and agricultural wastes potential

A variety of high energy crops have been investigated (Jönsson et al., 2013; Saini, Saini, & Tewari, 2015). These crops, along with microalgae, are considered as potential resources of renewable energy that can contribute toward the reduction of global dependence on fossil fuels. While these energy resources are naturally occurring and abundant without the being associated with environmental hazards, they are viable feedstocks for sustainable and clean fuel production (Musa, Zhonghua, Ibrahim, & Habib, 2018; Pattanaik et al., 2019). Consequently, the issue of technological policy advancement and the acceptance of the notion to install renewable energy systems or prepare lands for biofuel production and agrowaste beneficiation are easy decisions for developing countries desiring to exploit the benefits of clean energy production (Dahiya et al., 2018). However, most rural sub-Saharan African region cannot commit to the financial status quo required to meet this development (Onu & Mbohwa, 2018c).

Research has been conducted to group different agricultural feedstocks and potential agrowastes into different biofuel categories (proximate and ultimate analysis agricultural wastes) according to their evolution in terms of "generations" (Pattanaik et al., 2019). The biofuels (bioalcohols-bioethanol/biobutanol, biohydrogen, biogas, biooil, biogasoline, and biodiesel) classification includes the first, second, third, and fourth generations, depending on the characteristics of the biomass and the technology used for its conversion (Chen, 2014).

Composition and structure of lignocellulosic biomass

Lignocellulosic biomass (LCB) is an abundant, renewable source of carbohydrates, available for microbial conversion to chemicals and fuels. It is derived from agricultural wastes, which comprise of straw, wood, grass, and other agricultural discards (Domínguez-Bocanegra, Torres-Muñoz, & López, 2015). This type of biomass can be converted into liquid fuel. This, in turn, improves the CO_2 balance and, since it is a wasted resource, does not threaten the human food chain (Soccol et al., 2010). The composition of LCB is categorized into three main parts: cellulose (30%–50% dry wt.), hemicellulose (20%–40% dry wt.), and lignin (10%–20% dry wt.) (Putro et al., 2015). The molecular structures of cellulose, hemicellulose, and lignin are shown in Fig. 1.4A–C, respectively. Detailed insight into the composition and structure of the lignocellulosic biomass has been provided in Chapters 3 and 4.

Farm resource and food produce that are specifically known to have free sugar or starch content (e.g., sugar cane, and maize) in considerably large amount and vegetable oils and animal fats form the categories of feedstocks classified for the production of the first generation of biofuels. The unfortunate concern with regard to the first generation of biofuel biomasses is linked to their sensitive nature to changing climatic conditions, hence affecting their cultivation and availability. The feedstock can undergo rapid conversion and leads to high biofuel yield (Aro, 2016; DoE, 2011). The next generation of feedstock categorizes the second and the third

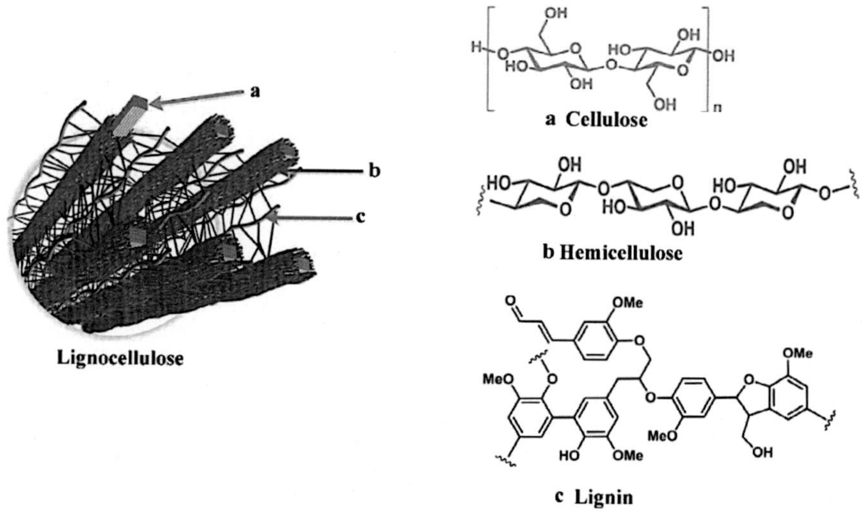

FIGURE 1.4

Lignocellulosic biomass: (A) cellulose; (B) hemicellulose; (C) lignin.

generations in one, due to the nature of their end product (bioethanol and biohydrogen). They are nonfood crops of very high starchy composition (biomass). The LCB resources are well discussed in subsequent chapters and are regarded as the second-generation biofuel. Also, while the third generation of biofuels are made from algae (biologically engineered crops), the fourth generation of biofuels entail advanced technological option that combines genetically optimized feedstocks, with the aims to capture and store carbon dioxide at the different stages of the production operation. The last two generations mentioned are still conceptualized; research is ongoing to contribute to its development to meet commercial-scale production (Onu & Mbohwa, 2019d).

Current disposition to commercialize biofuels production focuses on the first-generation feedstocks, while exploring the different resouces available to meet the ever-increasing global search for sustainable biofuels production (Jönsson et al., 2013; Taha et al., 2015). However, more attention is attributed to the second and third generations (high starchy, nonfood crops), as the conversion strategies and route toward biofuel production vary according to their classification.

1.4 Conclusions and perspectives

This chapter has identified waste management strategies that support sustainable agriculture diversification, tenable to mitigate capital-intensive implications of agricultural development in growing economies. Different waste management processes such as recovery, recycling, and reuse are discussed, giving inferences to technologies and viable crops that have the potential to significantly promote food and energy security. While the global food scarcity is significantly felt in Africa and other developing countries, leaving close to a billion people hungry (FAO, IFAD, & WFP, 2015), the world continues to generate tons (millions) of agricultural and food wastes yearly. According to research predictions; "The world population figure climb above 9 billion by 2050, and 11 billion in the subsequent century" (Food and Agriculture Organization of the United Nations, 2017). There is a great need to assimilate sustainability in line with agricultural development and intensify waste management operations or schemes that will manage trash collection, inspire product creation from agricultural refuse, and encourage the general industrial sector to function as a zero-waste contributor. The design therefore of an integrated sustainable waste management model that can accommodate different strategies to salvage the situation in the sub-Saharan African region is timely.

References

Achinas, S., & Euverink, G. J. W. (2016). Consolidated briefing of biochemical ethanol production from lignocellulosic biomass. *Electronic Journal of Biotechnology.* https://doi.org/10.1016/j.ejbt.2016.07.006.

AGRA. (2017). *Africa agriculture status report: The business of smallholder agriculture in sub-Saharan Africa.* AGRA. Retrieved from http://hdl.handle.net/10568/42343.

Agriculture in sub-Saharan Africa: Prospects and challenges for the next decade.(2016). https://doi.org/10.1787/agr_outlook-2016-5-en.

Akiyode, O., Tumushabe, A., Hadijjah, K., & Onu, P. (2017). Climate change, food security and environmental security: A conflict inclination assessment of Karamoja region of Uganda. *International Journal of Scientific World.* https://doi.org/10.14419/ijsw.v5i2.8458.

Amsterdam, H., & Thopil, G. A. (2017). Enablers towards establishing and growing South Africa's waste to electricity industry. *Waste Management.* https://doi.org/10.1016/j.wasman.2017.06.051.

Andriani, D., Wresta, A., Saepudin, A., & Prawara, B. (2015). A review of recycling of human excreta to energy through biogas generation: Indonesia case. *Energy Procedia, 68,* 219–225. https://doi.org/10.1016/j.egypro.2015.03.250.

Aro, E. M. (2016). From first generation biofuels to advanced solar biofuels. *Ambio.* https://doi.org/10.1007/s13280-015-0730-0.

Bajwa, D. S., Peterson, T., Sharma, N., Shojaeiarani, J., & Bajwa, S. G. (2018). A review of densified solid biomass for energy production. *Renewable and Sustainable Energy Reviews.* https://doi.org/10.1016/j.rser.2018.07.040.

Bing, X., Bloemhof, J. M., Ramos, T. R. P., Barbosa-Povoa, A. P., Wong, C. Y., & van der Vorst, J. G. A. J. (2016). Research challenges in municipal solid waste logistics management. *Waste Management.* https://doi.org/10.1016/j.wasman.2015.11.025.

Bond, A. J., & Morrison-Saunders, A. (2011). Re-evaluating sustainability assessment: Aligning the vision and the practice. *Environmental Impact Assessment Review.* https://doi.org/10.1016/j.eiar.2010.01.007.

Bruce, M.,C., James, H., Janie, R., Clare, M.,S., Stephen, T., & Eva, L. W. (2018). Urgent action to combat climate change and its impacts (SDG 13): Transforming agriculture and food systems. *Current Opinion in Environmental Sustainability.* https://doi.org/10.1016/j.cosust.2018.06.005.

Brundtland, G. H. (1987). *Our common future (Brundtland report).* United Nations Commission. https://doi.org/10.1080/07488008808408783.

Carrillo-Nieves, D., Rostro Alanís, M. J., de la Cruz Quiroz, R., Ruiz, H. A., Iqbal, H. M. N., & Parra-Saldívar, R. (2019). Current status and future trends of bioethanol production from agro-industrial wastes in Mexico. *Renewable and Sustainable Energy Reviews.* https://doi.org/10.1016/j.rser.2018.11.031.

Chen, H. (2014). *Biotechnology of lignocellulose.* https://doi.org/10.1007/978-94-007-6898-7.

Chifari, R., Lo Piano, S., Bukkens, S. G. F., & Giampietro, M. (2018). A holistic framework for the integrated assessment of urban waste management systems. *Ecological Indicators.* https://doi.org/10.1016/j.ecolind.2016.03.006.

Compton, M., Willis, S., Rezaie, B., & Humes, K. (2018). Food processing industry energy and water consumption in the Pacific northwest. *Innovative Food Science and Emerging Technologies.* https://doi.org/10.1016/j.ifset.2018.04.001.

Dahiya, S., Kumar, A. N., Shanthi Sravan, J., Chatterjee, S., Sarkar, O., & Mohan, S. V. (2018). Food waste biorefinery: Sustainable strategy for circular bioeconomy. *Bioresource Technology.* https://doi.org/10.1016/j.biortech.2017.07.176.

DoE. (2011). *Integrated resource plan for electricity 2010–2030.* Republic of South Africa: Department of Energy. https://doi.org/10.1016/j.wneu.2010.05.012.

Domínguez-Bocanegra, A. R., Torres-Muñoz, J. A., & López, R. A. (2015). Production of bioethanol from agro-industrial wastes. *Fuel, 149,* 85–89. https://doi.org/10.1016/j.fuel.2014.09.062.

Dong, L., Fujita, T., Dai, M., Geng, Y., Ren, J., Fujii, M., … Ohnishi, S. (2016). Towards preventative eco-industrial development: An industrial and urban symbiosis case in one typical industrial city in China. *Journal of Cleaner Production.* https://doi.org/10.1016/j.jclepro.2015.05.015.

Eisted, R., Larsen, A. W., & Christensen, T. H. (2009). Collection, transfer and transport of waste: Accounting of greenhouse gases and global warming contribution. *Waste Management and Research.* https://doi.org/10.1177/0734242X09347796.

Engg, M. (2013). Biogas as alternate fuel in diesel engines : A literature review. *IOSR Journal of Mechanical and Civil Engineering, 9*(1), 23–28.

FAO, IFAD, & WFP. (2015). *The state of food insecurity in the world. International hunger targets: Taking stock of uneven progress.* Retrieved from https://I4646E/1/05.15.

Ferranti, P. (2019). The United Nations sustainable development goals. In *Encyclopedia of food security and sustainability.* https://doi.org/10.1016/b978-0-08-100596-5.22063-5.

Food and Agriculture Organization of the United Nations. (2017). *SOFI 2017 – the state of food security and nutrition in the world.*

Glasziou, P., Altman, D. G., Bossuyt, P., Boutron, I., Clarke, M., Julious, S., … Wager, E. (2014). Reducing waste from incomplete or unusable reports of biomedical research. *The Lancet.* https://doi.org/10.1016/S0140-6736(13)62228-X.

Gomiero, T. (2018). Agriculture and degrowth: State of the art and assessment of organic and biotech-based agriculture from a degrowth perspective. *Journal of Cleaner Production.* https://doi.org/10.1016/j.jclepro.2017.03.237.

Guerrero, L. A., Maas, G., & Hogland, W. (2013). Solid waste management challenges for cities in developing countries. *Waste Management, 33*(1), 220–232. https://doi.org/10.1016/j.wasman.2012.09.008.

Gutiérrez, M. C., Serrano, A., Siles, J. A., Chica, A. F., & Martín, M. A. (2017). Centralized management of sewage sludge and agro-industrial waste through co-composting. *Journal of Environmental Management, 196*, 387–393. https://doi.org/10.1016/j.jenvman.2017.03.042.

Ikhlayel, M. (2018). An integrated approach to establish e-waste management systems for developing countries. *Journal of Cleaner Production, 170*, 119–130. https://doi.org/10.1016/j.jclepro.2017.09.137.

Ikumapayi, O. M., Akinlabi, E. T., Onu, P., Akinlabi, S. A., & Agarana, M. C. (2019). A generalized model for automation cost estimating systems (ACES) for sustainable manufacturing. *Journal of Physics: Conference Series, 1378*(3). https://doi.org/10.1088/1742-6596/1378/3/032043.

International Energy Agency. (2016). *World energy outlook 2016*. International Energy Agency. Retrieved from http://www.iea.org/publications/freepublications/publication/WEB_WorldEnergyOutlook2015ExecutiveSummaryEnglishFinal.pdf.

Jones, F., & Schoonover, R. (2010). Glossary of statistical terms. In *Handbook of mass measurement*. https://doi.org/10.1201/9781420038453.ch6.

Jönsson, L. J., Alriksson, B., & Nilvebrant, N. O. (2013). Bioconversion of lignocellulose: Inhibitors and detoxification. *Biotechnology for Biofuels*. https://doi.org/10.1186/1754-6834-6-16.

Kumar, S., Dhingra, A. K., & Singh, B. (2018). Process improvement through lean-Kaizen using value stream map: A case study in India. *International Journal of Advanced Manufacturing Technology*. https://doi.org/10.1007/s00170-018-1684-8.

Lawal, A. Q. T., Ninsiima, E., Odebiyib, O. S., Hassan, A. S., Oyagbola, I. A., Onu, P., & .Yusuf, A. D. (2019). Effect of unburnt rice husk on the properties of concrete. *Procedia Manufacturing, 35*, 635–640. https://doi.org/10.1016/j.promfg.2019.06.006.

Lebreton, L., & Andrady, A. (2019). Future scenarios of global plastic waste generation and disposal. *Palgrave Communications*. https://doi.org/10.1057/s41599-018-0212-7.

Leichenko, R. (2011). Climate change and urban resilience. *Current Opinion in Environmental Sustainability*. https://doi.org/10.1016/j.cosust.2010.12.014.

Liebetrau, J., Sträuber, H., Kretzschmar, J., Denysenko, V., & Nelles, M. (2019). Anaerobic digestion. *Advances in Biochemical Engineering/Biotechnology*. https://doi.org/10.1007/10_2016_67.

Liuzzi, S., Sanarica, S., & Stefanizzi, P. (2017). Use of agro-wastes in building materials in the Mediterranean area: A review. *Energy Procedia, 126*, 242–249. https://doi.org/10.1016/j.egypro.2017.08.147.

Long, T. B., Blok, V., & Coninx, I. (2016). Barriers to the adoption and diffusion of technological innovations for climate-smart agriculture in Europe: Evidence from the Netherlands, France, Switzerland and Italy. *Journal of Cleaner Production*. https://doi.org/10.1016/j.jclepro.2015.06.044.

Lynd, L. R., Liang, X., Biddy, M. J., Allee, A., Cai, H., Foust, T., … Wyman, C. E. (2017). Cellulosic ethanol: Status and innovation. *Current Opinion in Biotechnology*. https://doi.org/10.1016/j.copbio.2017.03.008.

Manyuchi, M. M., Mbohwa, C., & Muzenda, E. (2017). *Resource recovery from municipal waste and bio solids (digestate) through vermicomposting : A waste management initiative* (pp. 2–6).

Mao, C., Feng, Y., Wang, X., & Ren, G. (2015). Review on research achievements of biogas from anaerobic digestion. *Renewable and Sustainable Energy Reviews.* https://doi.org/10.1016/j.rser.2015.02.032.

Meena, M. D., Yadav, R. K., Narjary, B., Yadav, G., Jat, H. S., Sheoran, P., … Moharana, P. C. (2019). Municipal solid waste (MSW): Strategies to improve salt affected soil sustainability: A review. *Waste Management.* https://doi.org/10.1016/j.wasman.2018.11.020.

Meneses-Jácome, A., Diaz-Chavez, R., Velásquez-Arredondo, H. I., Cárdenas-Chávez, D. L., Parra, R., & Ruiz-Colorado, A. A. (2016). Sustainable Energy from agro-industrial wastewaters in Latin-America. *Renewable and Sustainable Energy Reviews.* https://doi.org/10.1016/j.rser.2015.12.036.

Mooney, H., Larigauderie, A., Cesario, M., Elmquist, T., Hoegh-Guldberg, O., Lavorel, S., … Yahara, T. (2009). Biodiversity, climate change, and ecosystem services. *Current Opinion in Environmental Sustainability.* https://doi.org/10.1016/j.cosust.2009.07.006.

Musa, S. D., Zhonghua, T., Ibrahim, A. O., & Habib, M. (2018). China's energy status: A critical look at fossils and renewable options. *Renewable and Sustainable Energy Reviews, 81*(June 2017), 2281–2290. https://doi.org/10.1016/j.rser.2017.06.036.

Ngoc, U. N., & Schnitzer, H. (2008). *Waste management towards zero emissions approach in the fruit juice processing industry* (pp. 91–97).

Nwakaire, J. N., Ezeoha, S. L., & Ugwuishiwu, B. O. (2013). Production of cellulosic ethanol from wood sawdust. *Agricultural Engineering International: CIGR Journal.*

Onu, P., Abolarin, M. S., & Anafi, F. O. (2017). Assessment of effect of rice husk ash on burnt properties of badeggi clay. *International Journal of Advanced Research, 5*(5), 240–247. https://doi.org/10.21474/IJAR01/4103.

Onu, P., & Mbohwa, C. (2018a). Future energy systems and sustainable emission control: Africa in perspective. *Proceedings of the International Conference on Industrial Engineering and Operations Management.*

Onu, P., & Mbohwa, C. (2018b). Green supply chain management and sustainable industrial Practices: Bridging the gap. *Proceedings of the International Conference on Industrial Engineering and Operations Management*, 786–792 (Washington DC).

Onu, P., & Mbohwa, C. (2018c). The interlink between sustainable supply chain management and technology development in industry. *Proceedings of the International Conference on Industrial Engineering and Operations Management*, 425–430 (Pretoria/Johannesburg).

Onu, P., & Mbohwa, C. (2019a). Sustainable supply chain management: Impact of practice on manufacturing and industry development. *Journal of Physics: Conference Series, 1378*(2). https://doi.org/10.1088/1742-6596/1378/2/022073.

Onu, P., & Mbohwa, C. (2019b). *Advances in solar photovoltaic grid parity.* https://doi.org/10.1109/IRSEC48032.2019.9078175.

Onu, P., & Mbohwa, C. (2019c). *Concentrated solar power technology and thermal energy Storage : A brief overview of nascent sustainable designs.* https://doi.org/10.1109/IRSEC48032.2019.9078156.

Onu, P., & Mbohwa, C. (2019d). Renewable energy technologies in brief. *International Journal of Scientific & Technology Research, 8*(10), 1283–1289.

Onu, P., & Mbohwa, C. (2019e). A sustainable industrial development approach: Enterprise risk management in view. In *IOP conference series: Materials science and engineering.*

Onu, P., & Mbohwa, C. (2019f). *Cloud computing and IOT application : Current statuses and prospect for industrial development* (pp. 1–14).

Onu, P., & Mbohwa, C. (2019g). Industrial energy conservation initiative and prospect for sustainable manufacturing. *Procedia Manufacturing, 35*, 546–551. https://doi.org/10.1016/j.promfg.2019.05.077.

Onu, P., & Mbohwa, C. (2019h). *New future for sustainability and industrial Development : Success in blockchain , Internet of production, and cloud computing technology* (pp. 1–12).

Onu, P., & Mbohwa, C. (2020a). *Industry 4.0 opportunities in manufacturing SMEs: Sustainability outlook*. Ota: ICESW.

Onu, P., & Mbohwa, C. (2020b). *Reimagining the future: Techno innovation advancement in manufacturing*. Ota: ICESW.

Onu, P., & Mbohwa, C. (2020c). *Green strategies and techno-innovation penetration: Sustainability advancement and manufacturing sector in perspective*. Muldersdrift: SAIIE.

Onu, P., & Mbohwa, C. (2020d). *On achieving biodiversity and the sustainable development goal – reviewing Africa's role*. Muldersdrift: SAIIE.

O'Sullivan, C. A., Bonnett, G. D., McIntyre, C. L., Hochman, Z., & Wasson, A. P. (2019). Strategies to improve the productivity, product diversity and profitability of urban agriculture. *Agricultural Systems*. https://doi.org/10.1016/j.agsy.2019.05.007.

Panesar, R., Kaur, S., & Panesar, P. S. (2015). Production of microbial pigments utilizing agro-industrial waste: A review. *Current Opinion in Food Science*. https://doi.org/10.1016/j.cofs.2014.12.002.

Pattanaik, L., Pattnaik, F., Saxena, D. K., & Naik, S. N. (2019). Biofuels from agricultural wastes. In *Second and third generation of feedstocks*. https://doi.org/10.1016/b978-0-12-815162-4.00005-7.

Pierpaoli, E., Carli, G., Pignatti, E., & Canavari, M. (2013). Drivers of precision agriculture technologies adoption: A literature review. *Procedia Technology*. https://doi.org/10.1016/j.protcy.2013.11.010.

Pivoto, D., Waquil, P. D., Talamini, E., Finocchio, C. P. S., Dalla Corte, V. F., & de Vargas Mores, G. (2018). Scientific development of smart farming technologies and their application in Brazil. *Information Processing in Agriculture*. https://doi.org/10.1016/j.inpa.2017.12.002.

Pollet, B. G., Staffell, I., & Adamson, K.-A. (2015). Current energy landscape in the Republic of South Africa. *International Journal of Hydrogen Energy, 40*(46), 16685–16701. https://doi.org/10.1016/j.ijhydene.2015.09.141.

Pradhan, A., & Mbohwa, C. (2014). Development of biofuels in South Africa: Challenges and opportunities. *Renewable and Sustainable Energy Reviews*. https://doi.org/10.1016/j.rser.2014.07.131.

Prakash, S. B. N. (2011). *Asian Journal of Food and Agro-Industry, 4*(01), 31–46.

Prapinagsorn, W., Sittijunda, S., & Reungsang, A. (2017). Co-digestion of napier grass and its silage with cow dung for methane production. *Energies*. https://doi.org/10.3390/en10101654.

Putro, J. N., Kurniawan, A., Soetaredjo, F. E., Lin, S. Y., Ju, Y. H., & Ismadji, S. (2015). Production of gamma-valerolactone from sugarcane bagasse over TiO_2-supported platinum and acid-activated bentonite as a co-catalyst. *RSC Advances*. https://doi.org/10.1039/c5ra06180f.

Qu, D., Wang, X., Kang, C., & Liu, Y. (2018). Promoting agricultural and rural modernization through application of information and communication technologies in China. *International Journal of Agricultural and Biological Engineering*. https://doi.org/10.25165/j.ijabe.20181106.4228.

Raji, I. O., & Onu, P. (2017). Untapped wealth potential in fruit for Uganda community. *International Journal of Advanced Academic Research, 3*(February), 17–25. Retrieved from http://www.ijaar.org/articles/Volume3-Number1/Sciences-Technology-Engineering/ijaar-ste-v3n1-jan17-p7.pdf.

Rothaermel, F. T., & Hess, A. M. (2010). Innovation strategies combined. *MIT Sloan Management Review.*

Saikia, D. (2011). Trends in agriculture-industry interlinkages in India: Pre and post-reform scenario. In *SSRN.* https://doi.org/10.2139/ssrn.1858203.

Saini, J. K., Saini, R., & Tewari, L. (2015). Lignocellulosic agriculture wastes as biomass feedstocks for second-generation bioethanol production: Concepts and recent developments. *3 Biotech.* https://doi.org/10.1007/s13205-014-0246-5.

Scarlat, N., Motola, V., Dallemand, J. F., Monforti-Ferrario, F., & Mofor, L. (2015). Evaluation of energy potential of municipal solid waste from African urban areas. *Renewable and Sustainable Energy Reviews, 50,* 1269–1286. https://doi.org/10.1016/j.rser.2015.05.067.

Seadon, J. K. (2010). Sustainable waste management systems. *Journal of Cleaner Production.* https://doi.org/10.1016/j.jclepro.2010.07.009.

Sheahan, M., & Barrett, C. B. (2017). Ten striking facts about agricultural input use in sub-Saharan Africa. *Food Policy.* https://doi.org/10.1016/j.foodpol.2016.09.010.

Shedrack, G. M., Yawas, D. S., Emmanuel, S. G., Peter, O., Dagwa, I. M., & Ibrahim, G. Z. (2019). Effect of postweld heat treatment on the mechanical behaviour of austinitic stainless steel. *Global Journal of Engineering Science and Research Management, 6*(4), 1–9. https://doi.org/10.5281/zenodo.2639252.

Skaggs, R. L., Coleman, A. M., Seiple, T. E., & Milbrandt, A. R. (2018). Waste-to-energy biofuel production potential for selected feedstocks in the conterminous United States. *Renewable and Sustainable Energy Reviews, 82*(October 2017), 2640–2651. https://doi.org/10.1016/j.rser.2017.09.107.

Soccol, C. R., Vandenberghe, L. P. de S., Medeiros, A. B. P., Karp, S. G., Buckeridge, M., Ramos, L. P., … Torres, F. A. G. (2010). Bioethanol from lignocelluloses: Status and perspectives in Brazil. *Bioresource Technology.* https://doi.org/10.1016/j.biortech.2009.11.067.

Song, M., Fisher, R., & Kwoh, Y. (2019). Technological challenges of green innovation and sustainable resource management with large scale data. *Technological Forecasting and Social Change.* https://doi.org/10.1016/j.techfore.2018.07.055.

Song, Q., Wang, Z., Li, J., & Duan, H. (2012). Sustainability evaluation of an e-waste treatment enterprise based on emergy analysis in China. *Ecological Engineering.* https://doi.org/10.1016/j.ecoleng.2012.02.016.

Taha, M., Shahsavari, E., Al-Hothaly, K., Mouradov, A., Smith, A. T., Ball, A. S., & Adetutu, E. M. (2015). Enhanced biological straw saccharification through coculturing of lignocellulose-degrading Microorganisms. *Applied Biochemistry and Biotechnology.* https://doi.org/10.1007/s12010-015-1539-9.

Tepic, M., Fortuin, F., Kemp, R. G. M., & Omta, O. (2014). Innovation capabilities in food and beverages and technology -based innovation projects. *British Food Journal.* https://doi.org/10.1108/BFJ-09-2011-0243.

United Nations (UN). (2015). Sustainable development goals. *Knowledge Platform.*

Woldeyohannes, A. D., Woldemichael, D. E., & Baheta, A. T. (2016). Sustainable renewable energy resources utilization in rural areas. *Renewable and Sustainable Energy Reviews, 66,* 1–9. https://doi.org/10.1016/j.rser.2016.07.013.

Xu, W., Fu, S., Yang, Z., Lu, J., & Guo, R. (2018). Improved methane production from corn straw by microaerobic pretreatment with a pure bacteria system. *Bioresource Technology.* https://doi.org/10.1016/j.biortech.2018.02.046.

Yahya, N. (2018). Agricultural 4.0: Its implementation toward future sustainability. In *Green energy and technology.* https://doi.org/10.1007/978-981-10-7578-0_5.

Yin, Y., Stecke, K. E., & Li, D. (2018). The evolution of production systems from Industry 2.0 through Industry 4.0. *International Journal of Production Research.* https://doi.org/10.1080/00207543.2017.1403664.

Yusuf, A. A., Peter, O., Hassanb, A. S., Tunji, A.,L., Oyagbola, I. A., Mustafad, M. M., & Yusuf, D. A. (2019). Municipality solid waste management system for Mukono district. *Procedia Manufacturing, 35*, 613–622. https://doi.org/10.1016/j.promfg.2019.06.003.

Zabed, H., Sahu, J. N., Boyce, A. N., & Faruq, G. (2016). Fuel ethanol production from lignocellulosic biomass: An overview on feedstocks and technological approaches. *Renewable and Sustainable Energy Reviews.* https://doi.org/10.1016/j.rser.2016.08.038.

CHAPTER

Agricultural wastes and opportunities in the food production chain

2

2.1 Introduction

One of the United Nations' sustainable development goals, which focuses on the elimination of hunger, does so through "tight" food security campaigns. Development in agricultural activities is critical for sustainable livelihood and ecosystem conservation. The food processing industry is faced with the challenge of improving its business exploits as part of its role toward the enrichment of several countries' economies (Compton, Willis, Rezaie, & Humes, 2018). Regardless, the waste generation level is dependent on their capacity to effectively manage respective agricultural or food processing operation, with consideration given mostly to waste prevention, and emphasis on zero harmful emissions among other perspectives that are subsequently addressed (Gustavsson, Cederberg, Sonesson, van Otterdijk, & Meybeck, 2011; Mukama et al., 2016). Moreover, the economics of scale for operations that support sustainable development is expected to overcome several technicalities to transform wastes into wealth. In essence, exploitation of animal feedstock and by-products form essential organic waste generation channels for clean energy production is advised.

New and developing trends in sustainable fuel production to decarbonize future combustion operations present an opportunity for the expansion of agricultural waste conversion technologies (Onu & Mbohwa, 2018a). The socioeconomic and environmental benefits of agrowaste diversity are seen in a closed-loop system of farming and recycling process, which prioritizes zero-waste generation, i.e., conservation of water, energy, land, and human resources. Agricultural activities must become a prerogative and attract advanced techniques (Ferranti, 2019; Food and Agriculture Organization of the United Nations, 2019). Thus, places demand more for effective operation, both for the purpose of meeting optimal food production and the cost—benefit of environmental protection (Abdulfatah Abdu Yusuf et al., 2019; Onu & Mbohwa, 2019b).

Agrowastes, as we know, have diverse potential, from the vast resources, which makes up the composition; domestic and industrial slurry/diary wastes, farm-crop residues and trim-offs, and several other categories present significant opportunities for sustainable intensification (Pattanaik, Pattnaik, Saxena, & Naik, 2019). Already,

Agricultural Waste Diversity and Sustainability Issues. https://doi.org/10.1016/B978-0-323-85402-3.00001-2

food security is estimated to increase beyond its current value by 60%, to meet the growing population value three decades from now. The above analogy is a worst-case scenario for sub-Saharan Africa and south-Asian/developing countries where the value is expected to increase above 132% by 2050 (FAO, 2017). This argument is true and poses that for efficient human feeding requirement, the demand for agricultural produce ought to meet zero waste production (Girotto, Alibardi, & Cossu, 2015). More so, these wastes being generated from direct farming/agricultural activities may be explored for different purposes and receive insight in this chapter. The authors consider organic waste from arable production returned to the farm as manure, and the lot of unharvested or discarded crop remains, often regarded as poor-quality, pest-infected, agriproduce as beneficial wastes. These wastes can be utilized as alternative resources in ecological conservation and clean energy advancements (Onu, Abolarin, & Anafi, 2017).

However, to a large extent, animal wastes coming from market-based animal farms are used to nitrify soils for farming activities. Wash-offs of these wastes/animal fertilizers that find their way to domestically useful water bodies leave potentially ill-causative effects that become harmful over time to humans and destroy aquatic life forms in the long run. The noxious gases due to the animal dung (nitrogenous content) converted to ammonia owing to their water-soluble nature find their way to the atmosphere, leading to acidic rains and carcinogenic diseases and further to the deposition of greenhouse gases, which contributes to global warming and climatic degradation effects (Dong et al., 2016). Over the years, investigations regarding the reduction of incessant dumping and disposal of agrowastes have led to the development of conversion techniques that capture the wastes and transform them into other forms (Despeisse, Mbaye, Ball, & Levers, 2012; Ngoc & Schnitzer, 2008). An anaerobic digestion arrangement is a system that is peculiar to facilitate organic wastes conversion for biogas production at small medium and large scale to meet both heat and electricity demand domestically or for commercial purposes. Anaerobic process has become more viable, with simple designs, cost performance, and efficient and effective operational capabilities, thus reducing the dependency on fossils (Mao, Feng, Wang, & Ren, 2015).

2.2 Correlation between agrowastes management and food production operation

Previously, the agriculture sector did not receive so much attention and considerations based on controls or mitigation of wastes disposal. These wastes were either hipped and burnt, allowed to dry and decay, or buried. Unfortunately, these practices are continued in most remote developing African localities. Nonetheless, waste quantification exercises and regulations that support agricultural waste management frameworks are being patronized and prioritized for future disposition, especially for income generation and as a sustainable energy generation option (Onu & Mbohwa, 2019c). National waste management and strategy for implementation in Africa began since early 2006. Although the initiative was first introduced in South Africa (Deat, 2005), it gradually gained popularity through the efforts of different agencies that subscribe to conservative approaches for farming, thereby furnishing the government in their reports. This was based on investigative studies to have a well-organized agrowaste management scheme. While new opportunities set in, regarding businesses within the agrowaste management framework, collection and transport, processing and treatment, and utilization before onward disposal, development of new regulation is contemplated for total agricultural waste management and prevention through the reuse and recycle approach mechanism.

The future looks to energy derivation from wastes, especially of agricultural and organic sources, in several ways (Komakech, Zurbrügg, Miito, Wanyama, & Vinnerås, 2016; UNFCC, 2015). The context of agroindustry residues draws emphasis on food processing industries, which opens a new path to sustainable product development that can mitigate the underlying challenges of incessant disposal of wastes to the environment. For example, bioextracts from agrowastes such as fruit peels can be used as aromatics. The marriage of waste management activities in the agricultural sector, with cleaner production initiatives, has been hauled in research. According to Hall & Howe, 2012, "the continuous application of an integrated, preventive, environmental strategy applied to processes, products, and services to increase overall efficiency and reduce risk to humans and the environment" (Hall & Howe, 2012) can cause change, especially with the perception about the little potential thought about concerning agrowaste diversity. Waste resource valorization from agriculture, food industries, and domestic holdings can contribute to the triple bottom line responsivity (Pleissner, Sze, & Lin, 2013; Waldron, 2009).

2.3 Waste management and implication of biodiversity: a brief into industrial food production

The conversion of agricultural waste biomass or feedstock, i.e., organic/cellulosic materials for biofuels production, is conducted through a process called biorefinery. Biogas, biohydrogen, bioethanol, biodiesel, biooils, and biochar, all form part of the constituent and benefits that can be derived from the conversion process of agrowaste (Hellsmark & Söderholm, 2017). This section captures an area of predominant agricultural waste occurrences, the activities that are tantamount to food, water, and energy wastage, and its effects on the general output, with regard to wastes from agricultural resources and during its exploitation. For example, the trend in biofuel development is projected to reach production of up to 159 billion liters before 2025 (OECD, 2018). Already, the transportation sector is committed to driving biofuel production, with a global expectation to meet up to 26% of the total demand by 2050 (International Bioenergy Agency, 2017).

The Republic of South Africa in her biowastes exploits is in front to apply its new corporate practice of greening, the city of Johannesburg's transport system, through the acquisition of hundreds of dual-fuel buses, to support zero harmful emission release in the transport sector. The Euro-V-DDF brand buses run on natural gas and diesel with an up to 80% substitution, thus promoting cost savings on diesel and a reduction in the environmental impact. The contribution aligns the use of biofuel technology and other initiatives as per electric and fuel cell buses to promote sustainable living in the city of Johannesburg—South Africa, thus forming part of the government initiative to advance climate change mitigation.

Food processing industries

The focus on utilization of agrofood waste has indeed attracted enormous research and nascent application in the food process industries, within the three geoselections (the west, east, and southern parts of Africa, sighting, Nigerian, Uganda, and the Republic of South Africa) as per the present study. So far, the practice of using the mechanically generated electricity to run the operations in a food processing firm is what is obtainable and contributes to the high cost of agrowaste management. A body of literature has explored the energy use and energy-saving potentials in the food processing industry in the past decade; the results show a trend of improvement in cost-effectiveness and reduction in industrial energy consumption through effective energy management practices and energy-efficient equipment brought about by the introduction of newer technologies and techniques (Attah, 2013; Hall & Howe, 2012; Mohammed, Mustafa, Bashir, & Ibrahem, 2017).

The idea is to have an all-encompassing and enriching process where agricultural resources, after leaving the farm, are transformed into valuable products with minimum wastes generation. The resolve is to cut energy consumption and reduce the implication on the environment, due to the emission effect of generating the electricity that powers the sector (IEA, 2017; Onu & Mbohwa, 2019c, Onu & Mbohwa, 2019a; Onu & Mbohwa, 2018). As such, efficient motors, boilers, compressors, furnaces, air-conditioning and space cooling, and industrial lighting appliances must be substituted by energy-efficient types, and the development and deployment of real-life biotechnologies can utilize the organic wastes in the food processing operation for both sustainable power generation and other potential gains is required (Ruane & Sonnino, 2011).

The food production industries in the sub-Saharan Africa are in dire need to transform their operational capabilities and also upgrade its organizational effectiveness to measure up to international standards in the verge to efficiently utilize the agricultural resources that come from the farms (Hepburn et al., 2019; Sharma, Sarkar, Singh, & Singh, 2017; Wanyama et al., 2016). The food and beverage industries, as well as the cereal/flour mills companies, breweries, meat, and dairy processing firm, not failing to mention the pharmaceuticals that exploit agricultural resources, all generate wastes. More importantly, the behavioral use of energy in the enterprises mentioned earlier is of central concern, as the sector (food processing industry) needs to consider energy and water-saving potential all through their operation (Ruane, Sonnino, & Agostini, 2010). The use of crude and traditional technologies by food production factories over the years in the areas of wastes reduction/recovery has gradually phased out and replaced by a list of anew. While the popular concept of waste management dwells on the assertion to feed the agricultural by-products to farm animals, the reality on-ground is that these animals can only feed as much of what is viable to them, and with the leftovers high in nitrates and cellulosic materials, must meet other resourcefulness. The aspect that deals with technological adoptions to treat waste reviews the usage of nonconventional technologies: ultrasounds and microwaves, high-voltage electrical discharge systems, and pressurized liquid extraction, and lists them as greener and relatively cheaper techniques (Muñoz, Meier, Diaz, & Jeison, 2015).

Furthermore, different technologies that can be applied to meet the desired minimum energy consumption throughout a product supply chain are expected to meet the sustainability criteria, which satisfies social, economic, and environmental performances (Onu & Mbohwa, 2019a, Onu & Mbohwa, 2019b; Onu & Mbohwa, 2019f). The conservation of energy, water, viable by-product resources, and controlled pollution from reckless effluents necessitate sustainable approaches and application of technological initiatives to support energy savings and cleaner production in the food processing/manufacturing industries, for example, reduction in freezing and heating time/capture of the heat (avoiding leakages) (Compton, Willis, Rezaie, & Humes, 2018; Dahiya et al., 2018).

The cleaner production and alternative energy development strategies, which aim to support sustainable development, and the global biodiversity framework relates to protecting nature and improve livelihood. As such, it is guided by policies that support the exploitation of biofuels for heat and electricity generation. Thus, the efficient utilization of agroresources and the processes tenable to lead value-added products calls for investment in sustainable strategies that look beyond traditional practices, and crude methods, ensuring proper training to propagate awareness in food waste management (Bodas Freitas, Dantas, & Iizuka, 2012). The consideration for the technology selection and suitability or the risk associated with any decision to regularize/minimize the energy dispensation in food process operation ought to align with government regularization and identifies or addresses the specific concerns of waste beneficiation and food wastage (Hall & Howe, 2012; Lin & Xie, 2015).

New challenges are springing up, regarding the nonavailability of lands for wastes disposal, and leave the rural farmers in the stride of negative environmental implications due to pollution. The food processing industry, therefore, can play an essential role in agricultural waste management and global food shortage challenges (Raji & Onu, 2017). The sector is bound to consider the prospect of waste management appraisal and midlevel worker's skill development to actualize the strategies for an effective agricultural waste treatment process. The criteria for the level of technology required to be used in the agricultural waste beneficiation process depend mostly on operational capacity and concentration of industries available to adopt the use of what technology for what specific reason and the ethical consideration for maintenance of the equipment purported for energy optimization, waste elimination, and in the long run, social and environmental gratification (Maertens & Barrett, 2013; Masebinu, Akinlabi, Muzenda, Mbohwa, & Aboyade, 2017).

Food waste management

The cycle of food production and waste management is interlinked with land conservation, energy reduction, water consumption, human resource control, and post-farming activities (Papargyropoulou, Lozano, Steinberger, Wright, & Ujang, 2014; Salomone et al., 2017). Recent investigations have identified high food waste at the point of final consumption and attribute high importance to consumers' behavior, based on their capacity to generate more wastes compared with any of the previously mentioned causatives, especially when the end products are discarded and left to waste (Gaiani, Caldeira, Adorno, Segrè, & Vittuari, 2018). Sound management practices have been proposed as a way to ensure that consumers, through informative workshops, training programs, and awareness campaigns are enlightened (Filimonau & Gherbin, 2018).

A gradual incentivization process that promotes innovation and developmental acceleration or improvement of waste reduction practices at farms before the final stage of product consumption is advised. Standardization, regularization, and implementation of green initiatives across the food production chain, along with investment for the adoption of reliable, cost-effective technology, will drive positive socioeconomic achievement and lead to societal satisfaction (OECD/IEA & IRENA, 2017). Close collaboration between the agricultural producers, research experts, the industry (food process), and other stakeholders is essential to evaluate and identify requirements and responsibilities of parties to collaborate and implement conservative strategies that will usher technical appreciation and behavioral change to promote waste management. The emerging technology trend that permits mobile applications and the use of internet services is fast rising, thus supporting smart monitoring and effective farming operations (Gondchawar & Kawitkar, 2016).

Energy activities in food process industries

The manufacturing sector is listed among the first three highest industry energy consumers, with the food processing industry contending for the highest share of electric/fuel and water consumption; it translates to increased operating costs, owing to the additional protocol to strategize for waste management (Food and Agriculture Organization of the United Nations, 2019). An energy intensity evaluation of food processing industries or the energy value of the agricultural products may not provide a holistic result without also identifying the measure of the energy in the different activities from the inception of cultivation. The pressure on the agricultural sector and the food processing industry due to the rising population growth rate, beyond 9 billion by 2050, have a ripple effect on energy and water consumption (UN-Water, 2016). The company may become overwhelmed by demand, which indirectly affects resources, thus requiring conservative approaches to necessitate reuse and recycling operations.

The food processing industries had always depended largely on grid electricity and energy from natural gas through the processing and utilization of their wastes for green energy potentials (Kang & Lee, 2016). A case where the food production company specializes in processing straight from the farm products; grinding/

crushing and milling operation tends to increase the energy consumption share, not to mention, other heating, cooling, and appropriate design for drying operations or industry ventilation process that contributes to energy consumption (Compton, Willis, Rezaie, & Humes, 2018; Skaggs, Coleman, Seiple, & Milbrandt, 2018). The consideration for burners, boilers, refrigeration, and fans replacement with new low frequency, less energy-consuming equipment, or as in the case of boilers with condensing economizers' is timely for energy cost saving. A situation of combined heat and power processes that have recorded remarkable results for meeting the food industry energy demand, pioneered by the United States, has achieved up to 40% capacity (Compton, Willis, Rezaie, & Humes, 2018).

The consideration for innovative tactics and nascent technology application to meet efficient recycling and reuse of water wastes, or reduction in energy consumption during agrowastes beneficiation, has been identified: (1) high-pressure processing of food, raw material, and product during sterilization and pasteurization purpose, (2) microwave heating process, which is more effective in combination with other drying techniques, (3) low-temperature moisture removal—ultrasound process, (4) refractance window dehydration, (5) pulsed electric field systems, achieved through rapid evaporation with reduced process time over most methods mentioned and of optimal performance in combined mode with the thermal process, (6) infrared: useful for food preservation and prevention of bacterial and fungus growth, (7) energy from industrial food wastewater, and the production of hydrogen to meet electricity demands, and (8) ozone technology.

Safety and quality assurance measures

Eminent disease-causative challenges affect the overall gains in agriculture activities and their waste management success, thus calling for food safety operations process. The food safety management system is a quality assurance program that approves the identification, evaluation, and control of hazards, which pose a threat to the food supply chain (Arvanitoyannis, 2009). Popular certification systems, which exist and currently adopted by numerous food processing firms, include Hazard Analysis and Critical Control Point (HACCP) for purpose of risk prevention/aversion due to physical, chemical, and microbiological hazards that occur during food processing (from agroresources, procurement, transfer/handling, manufacturing, distribution, to storage/consumption) and distribution. The ISO 22000:2005, which is a combination of the principles of quality manufacturing operations (ISO 9001:2001) and the HACCP principle (investigate the hazards, identify CCP, establish critical limits, monitor the CCP, advise about corrective actions, verify procedures, and record and document activities), is able to track food safety operations, relating to the interaction between two or more corporations, and activities within the firms' that make up the entire food supply chain, along with the safety measures to attain quality, service delivery, and customer health/product satisfaction.

Medical and life science development

Researchers have exploited the petrochemical content in some fruit peels for the production of specific drug by extracting the antioxidants from them

(Bobinaite et al., 2016). Fruit peels, leaves, and seeds of grapes, strawberry, apples, carrot pulp, and pineapple pomace are potential material sources for pharmacological, cosmetic, and dietic purposes and are found also to contain the compound, most of which may well play the role of natural preservatives which is essential in the food processing industries. More complex agrowastes like the Bagasse's from plants such as sugarcane and maize have also been identified to have a significant amount of phenolic and antioxidant (Zheng et al., 2017). In general, agricultural wastes from peels, leaves, and seeds find more exciting use in seasoning and manufacture of dyes and flavors, among other rich essential oils or chemical and nutrition that promote healthy living.

A brief on green energy and conversion techniques

An insight into agrowastes composition, for both commercial and small-scale business purposes, continues to exploit lignocellulose materials (biomass) and the component structures that form them, such as protein, organic acids, ash, lignin, and other non-structural materials, mostly identified among cash crops such as maize, rice, and from the bagasse's, cobs, and other postharvest residues. Based on the nonedible nature of these wastes and the excessive pressure which it leaves on land, irrespective of its potential for farm manure, it has also inspired exploitation for clean energy production. The complex intermediate processes involved in agrowaste conversion pose several levels of challenges in biorefinery development and frustrate its commercialization, thus limiting the production of biogas, bioethanol, biodiesel, and other value-added biochemicals (Balat, 2011; Lynd et al., 2017; Maurya, Singla, & Negi, 2015).

The process of anaerobic digestion is one in which microorganisms act upon, and breakdown biodegradable materials in the absence of oxygen (controlled aeration and temperature) to generate useful biogas (i.e., methane and/or carbon dioxide) (Montalvo, Vielma, Borja, Huiliñir, & Guerrero, 2018; Xu, Fu, Yang, Lu, & Guo, 2018). The process takes place under controlled operational temperature conditions for the gases to be produced: low, medium and higher temperature ranges, or psychrophilic (5–15°C), mesophilic (25–45°C), and the thermophilic (55–70°C) arrangement, respectively, having a varying level of performance and cost advantages (Elias, Wieczorek, Rosenne, & Tawfik, 2014; Jang, Choi, & Kan, 2018). The nature of energy recovered from the "digestive process" is such that it can be utilized for heating and electricity generation, depending on the capacity of the plant. Additionally, the biofuel produced can be developed to serve as fuels in both domestic and commercial transport systems that operate in cities and municipalities. This has already been noticed on a small-scale testing, in different cities: Morocco, Abuja-Nigeria, Johannesburg, and Accra-Ghana to list a few (Woldeyohannes, Woldemichael, & Baheta, 2016).

The viability of the anaerobic technology in agrowaste management is important for future energy systems development to substitute fossil fuels and reduce the pollution caused by burning the unwanted wastes (Soccol et al., 2010). The fermentation process is the occasion where conscious efforts are made to introduce yeasts and other preferable and available microorganisms to produce ethanol (liquid). The ethanolic produced is proceeded by conversion of the organic wastes to sugar (hydrolysis) before fermentation, and the final process of distillation takes place to obtain an ethyl-alcoholic nature of bio-based fuel. The traditional burning (sufficient oxygen supply) by the incineration process is posed as a nonsustainable method of handling disposed wastes, especially when there are no heat energy recovery capabilities. Interestingly, the thermochemical process of wastes gasification (under controlled oxygen supply) is gaining popularity in the production of syngas—hydrogen, carbon monoxide, and carbon dioxide (Dong et al., 2018). The pressurized gases produced under controlled pollution procedure may be trapped/collected and are sufficient to meet electricity generation, whereas the heat energy during the incineration and/or gasification processes can also be recovered in more modern systems and used for drying and most commonly for water steaming.

The pyrolysis process involves the thermochemical conversion and decomposition of organic/biodegradable contents of wastes at a very high temperature (usually above 450°) in an inert atmosphere that occurs in the absence of oxygen (Bundhoo, 2018). The pyrolysis activity causes an irreversible change in the chemical composition and physical state of the compounds with oils and gas fuels as the most beneficial end products of the process. Although not all technologies are suitable for the conversion of all categories of wastes, as sorting may become painstaking and result in high cost and time-wasteful operation. Notwithstanding, the discomfort due to irritating fumes from landfills and the greenhouse effect will be significantly reduced as well; the overreliance on fossils would decrease (Dhyani & Bhaskar, 2018).

Challenges and prospects of wastes management linked to food processing

Investments in waste management over the years have been stunted by high initial capital cost to execute, i.e., to acquire and operate the necessary technologies, as against the backdrop of high operating costs (Wang, Pu, Ragauskas, & Yang, 2019). Upon the consideration of the former, which accesses social and environment tendencies or when compared with the latter, to settle the challenges due to unstable fossil fuel prices/scarcity, an investment in bioenergy that supports sustainability becomes a sensible option. A look at the cost of conducting feasibility, including the cost of setting up the financial structure to fund the project, negotiation, supervision, and technical overview, or the cost of certification/registration/compliance and permit services, draws an inference to the economics of waste management. Other challenges that may hinder wastes management prospect include revenue acquisition for the agrowaste management operations, looking at the overall quality of the output from the processes, availability of the feedstocks, technology, government regulations, market and distance to the market, especially from the sub-Saharan Africa perspective (Carrillo-Nieves et al., 2019; Gebreegziabher, Naik, Melamu, & Balana, 2014).

Consequently, the era of the fourth industrial revolution seeks the most readily available mechanism to integrate digitization and smart operations in diverse fields and agricultural exploration (Onu & Mbohwa, 2019g; Puyol et al., 2017). The question goes to say, what measure in terms of authority of enforcement, monitoring, or perhaps how much of government incentive is considered as the adequate to decide technology or instruction to promote sustainability? The role of technology, government intervention, ethical compliance, and social supports are meant to facilitate an effective process of clean energy production from agricultural waste resources without necessarily having an impact in the supply chain of these resources (Guzman & Clapp, 2017; Newell, Pizer, & Raimi, 2013; Onu & Mbohwa, 2018b). While some researchers proffer that biofuel development from agrowaste may threaten food supply in animal husbandry, the argument is maintained on the basis that only the excesses of what is considered to waste (discards and no longer palatable) are processed to become green products. This will boost the economic concern of a decentralized energy solution that is free of environmental damage, made possible by the exploitation of organic and unwanted materials, thus holding promise for the sustainable green future envisaged.

2.4 Conclusions and perspectives

The need to assimilate the potential of the various wastes and processes optimization for value creation, and beneficial end products, ranging from biofuels, chemicals, sustainable materials, and a variety of spin-offs, and eco-by-products useful for domestic, and other engineering purposes has been addressed in this study. This chapter details waste streams and contributes to knowledge, based on strategies to advance operational excellence in food wastes management. The authors discussed long-term strategy that combines waste utilization and lean practice in farming and food production with consideration to optimize the economy of resource handling, production, and eco-conservation in the African context to lead effective agrowaste management activities.

The limitation of the unknown due to agricultural waste being used as feedstocks and the shortcoming in that regard or the diversity in farming techniques to replace traditional methods calls for more investigation and support from the higher institute of education. This is important to provide an understanding of information communication technology access in agricultural services extension and crop topology. Furthermore, based on the crucial potential of conservative agricultural schemes to foster communal and national development, investments in nascent energy systems will help to increase the conservational practice and renewable energy supply for industrial and socioeconomic benefits. The expectation is that cost-effective resources management and energy conservative techniques can be integrated into food processing industries to maximize materials and energy consumption and mitigate the threat of food and water scarcity.

References

Arvanitoyannis, I. S. (2009). HACCP and ISO 22000: Application to foods of animal origin. In *HACCP and ISO 22000: Application to Foods of Animal Origin*. (pp. 1–560). Wiley-Blackwell.

Attah, N. E. (2013). Possession by dispossession. *Journal of Land and Rural Studies, 1*(2), 213–228. https://doi.org/10.1177/2321024913513385.

Balat, M. (2011). Production of bioethanol from lignocellulosic materials via the biochemical pathway: A review. *Energy Conversion and Management, 52*(2), 858–875. https://doi.org/10.1016/j.enconman.2010.08.013.

Bobinaite, R., Viskelis, P., Bobinas, Č., Mieželiene, A., Alenčikiene, G., & Venskutonis, P. R. (2016). Raspberry marc extracts increase antioxidative potential, ellagic acid, ellagitannin and anthocyanin concentrations in fruit purees. *Lebensmittel-Wissenschaft und -Technologie- Food Science and Technology, 66*, 460–467. https://doi.org/10.1016/j.lwt.2015.10.069.

Bodas Freitas, I. M., Dantas, E., & Iizuka, M. (2012). The Kyoto mechanisms and the diffusion of renewable energy technologies in the BRICS. *Energy Policy, 42*, 118–128. https://doi.org/10.1016/j.enpol.2011.11.055.

Bundhoo, Z. M. A. (2018). Microwave-assisted conversion of biomass and waste materials to biofuels. *Renewable and Sustainable Energy Reviews, 82*, 1149–1177. https://doi.org/10.1016/j.rser.2017.09.066.

Carrillo-Nieves, D., Rostro Alanís, M. J., de la Cruz Quiroz, R., Ruiz, H. A., Iqbal, H. M. N., & Parra-Saldívar, R. (2019). Current status and future trends of bioethanol production from agro-industrial wastes in Mexico. *Renewable and Sustainable Energy Reviews, 102*, 63–74. https://doi.org/10.1016/j.rser.2018.11.031.

Compton, M., Willis, S., Rezaie, B., & Humes, K. (2018). Food processing industry energy and water consumption in the Pacific northwest. *Innovative Food Science & Emerging Technologies, 47*, 371–383. https://doi.org/10.1016/j.ifset.2018.04.001.

Dahiya, S., Kumar, A. N., Shanthi Sravan, J., Chatterjee, S., Sarkar, O., & Mohan, S. V. (2018). Food waste biorefinery: Sustainable strategy for circular bioeconomy. *Bioresource Technology, 284*, 2–12. https://doi.org/10.1016/j.biortech.2017.07.176.

Deat. (2005). National waste management strategy implementation South Africa recycling waste stream analysis and prioritisation for recycling. *Analysis*, 12/ 9/ 6.

Despeisse, M., Mbaye, F., Ball, P. D., & Levers, A. (2012). The emergence of sustainable manufacturing practices. *Production Planning & Control*, 354–376. https://doi.org/10.1080/09537287.2011.555425.

Dhyani, V., & Bhaskar, T. (2018). A comprehensive review on the pyrolysis of lignocellulosic biomass. *Renewable Energy, 129*, 695–716. https://doi.org/10.1016/j.renene.2017.04.035.

Dong, L., Fujita, T., Dai, M., Geng, Y., Ren, J., Fujii, M., … Ohnishi, S. (2016). Towards preventative eco-industrial development: An industrial and urban symbiosis case in one typical industrial city in China. *Journal of Cleaner Production*. https://doi.org/10.1016/j.jclepro.2015.05.015.

Dong, J., Tang, Y., Nzihou, A., Chi, Y., Weiss-Hortala, E., & Ni, M. (2018). Life cycle assessment of pyrolysis, gasification and incineration waste-to-energy technologies: Theoretical analysis and case study of commercial plants. *The Science of the Total Environment, 626*, 744–753. https://doi.org/10.1016/j.scitotenv.2018.01.151.

Elias, M., Wieczorek, G., Rosenne, S., & Tawfik, D. S. (2014). The universality of enzymatic rate-temperature dependency. *Trends in Biochemical Sciences, 39*, 1–7. https://doi.org/10.1016/j.tibs.2013.11.001.

FAO. (2017). The future of food and agriculture: Trends and challenges. In Graziano da Silva José (Ed.), *Food and agriculture organization of the United Nations*. Food and Agriculture Organization of the United Nations (FAO).

Ferranti, P. (2019). The United nations sustainable development goals. In United nations (Ed.), *Encyclopedia of food security and sustainability*. United nations.

Filimonau, V., & Gherbin, A. (2018). An exploratory study of food waste management practices in the UK grocery retail sector. *Journal of Cleaner Production, 167*, 1184–1195. https://doi.org/10.1016/j.jclepro.2017.07.229.

Food and Agriculture Organization of the United Nations. (2019). *Save food: Global initiative on food loss and waste reduction*. FAO.

Gaiani, S., Caldeira, S., Adorno, V., Segrè, A., & Vittuari, M. (2018). Food wasters: Profiling consumers' attitude to waste food in Italy. *Waste Management, 72*, 17–24. https://doi.org/10.1016/j.wasman.2017.11.012.

Gebreegziabher, Z., Naik, L., Melamu, R., & Balana, B. B. (2014). Prospects and challenges for urban application of biogas installations in Sub-Saharan Africa. *Biomass and Bioenergy, 70*, 130–140. https://doi.org/10.1016/j.biombioe.2014.02.036.

Girotto, F., Alibardi, L., & Cossu, R. (2015). Food waste generation and industrial uses: A review. *Waste Management, 45*, 32–41.

Gondchawar, N., & Kawitkar, P. R. S. (2016). IoT based smart agriculture. *International Journal of Advanced Research in Computer and Communication Engineering, 5*(6). https://doi.org/10.17148/IJARCCE.2016.56188.

Gustavsson, J., Cederberg, C., Sonesson, U., van Otterdijk, R., & Meybeck, A. (2011). Global food losses and food waste: Extent, causes and prevention. *International Congress: Save Food!*. https://doi.org/10.1098/rstb.2010.0126.

Guzman, L. I., & Clapp, A. (2017). Applying personal carbon trading: A proposed "carbon, health and savings system" for British Columbia, Canada. *Climate Policy, 17*(5). https://doi.org/10.1080/14693062.2016.1152947.

Hall, G. M., & Howe, J. (2012). Energy from waste and the food processing industry. *Process Safety and Environmental Protection, 90*(3), 203–212. https://doi.org/10.1016/j.psep.2011.09.005.

Hellsmark, H., & Söderholm, P. (2017). Innovation policies for advanced biorefinery development: Key considerations and lessons from Sweden. *Biofuels, Bioproducts and Biorefining, 11*, 28–40. https://doi.org/10.1002/bbb.1732.

Hepburn, C., Adlen, E., Beddington, J., Carter, E. A., Fuss, S., Mac Dowell, N., ... Williams, C. K. (2019). The technological and economic prospects for CO_2 utilization and removal. *Nature*. https://doi.org/10.1038/s41586-019-1681-6.

IEA. (2017). World energy balance: An overview of global trends (p. 21). *WORLD ENERGY BALANCES: AN OVERVIEW Global Trends* https://www.iea.org/publications/freepublications/publication/world-energy-balances—2017-edition—overview.html.

International Bioenergy Agency. (2017). *Technology roadmap: Delivering sustainable bioenergy*. International Energy Agency (IEA).

Jang, H. M., Choi, Y. K., & Kan, E. (2018). Effects of dairy manure-derived biochar on psychrophilic, mesophilic and thermophilic anaerobic digestions of dairy manure. *Bioresource Technology, 250*, 927–931. https://doi.org/10.1016/j.biortech.2017.11.074.

Kang, D., & Lee, D. H. (2016). Energy and environment efficiency of industry and its productivity effect. *Journal of Cleaner Production, 135*, 184–193. https://doi.org/10.1016/j.jclepro.2016.06.042.

Komakech, A. J., Zurbrügg, C., Miito, G. J., Wanyama, J., & Vinnerås, B. (2016). Environmental impact from vermicomposting of organic waste in Kampala, Uganda. *Journal of Environmental Management, 181*, 395–402. https://doi.org/10.1016/j.jenvman.2016.06.028.

Lin, B., & Xie, X. (2015). Factor substitution and rebound effect in China's food industry. *Energy Conversion and Management, 105*, 20–29. https://doi.org/10.1016/j.enconman.2015.07.039.

Lynd, L. R., Liang, X., Biddy, M. J., Allee, A., Cai, H., Foust, T., … Wyman, C. E. (2017). Cellulosic ethanol: Status and innovation. *Current Opinion in Biotechnology*. https://doi.org/10.1016/j.copbio.2017.03.008.

Maertens, A., & Barrett, C. B. (2013). Measuring social networks' effects on agricultural technology adoption. *American Journal of Agricultural Economics*, 353–359. https://doi.org/10.1093/ajae/aas049.

Mao, C., Feng, Y., Wang, X., & Ren, G. (2015). Review on research achievements of biogas from anaerobic digestion. *Renewable and Sustainable Energy Reviews, 45*, 540–555. https://doi.org/10.1016/j.rser.2015.02.032.

Masebinu, S. O., Akinlabi, E., Muzenda, E., Mbohwa, C., & Aboyade, A. O. (2017). Technoeconometric analysis of membrane technology for biogas upgrading. In *Transactions on engineering technologies: World congress on engineering and computer science 2015* (47th ed., 52, pp. 16929–16938). ACS.

Maurya, D. P., Singla, A., & Negi, S. (2015). An overview of key pretreatment processes for biological conversion of lignocellulosic biomass to bioethanol. *3 Biotech, 5*(5), 597–609. https://doi.org/10.1007/s13205-015-0279-4.

Mohammed, Y. S., Mustafa, M. W., Bashir, N., & Ibrahem, I. S. (2017). Existing and recommended renewable and sustainable energy development in Nigeria based on autonomous energy and microgrid technologies. *Renewable and Sustainable Energy Reviews, 75*, 820–838. https://doi.org/10.1016/j.rser.2016.11.062 (November 2016).

Montalvo, S., Vielma, S., Borja, R., Huiliñir, C., & Guerrero, L. (2018). Increase in biogas production in anaerobic sludge digestion by combining aerobic hydrolysis and addition of metallic wastes. *Renewable Energy, 123*, 541–548. https://doi.org/10.1016/j.renene.2018.02.004.

Mukama, T., Ndejjo, R., Musoke, D., Musinguzi, G., Halage, A. A., Carpenter, D. O., & Ssempebwa, J. C. (2016). Practices, concerns, and willingness to participate in solid waste management in two urban slums in Central Uganda. *Journal of Environmental and Public Health, 2016*, Article 6830163. https://doi.org/10.1155/2016/6830163.

Muñoz, R., Meier, L., Diaz, I., & Jeison, D. (2015). A review on the state-of-the-art of physical/chemical and biological technologies for biogas upgrading. *Reviews in Environmental Science and Biotechnology, 14*, 727–759. https://doi.org/10.1007/s11157-015-9379-1.

Newell, R. G., Pizer, W. A., & Raimi, D. (2013). Carbon markets 15 Years after kyoto: Lessons learned, new challenges. *The Journal of Economic Perspectives, 27*, 123–146. https://doi.org/10.1257/jep.27.1.123.

Ngoc, U. N., & Schnitzer, H. (2008). *Waste management towards zero emissions approach in the fruit juice processing industry* (pp. 91–97). WSEAS.

OECD. (2018). *Renewables information*. OECD.

OECD/IEA, & IRENA. (2017). Perspectives for the energy transition: Investment needs for a low-carbon energy system. In *International energy agency*. International Renewable Energy Agency. IRENA.

Onu, P., & Mbohwa, C. (2018a). Future energy systems and sustainable emission control: Africa in perspective. In *Proceedings of the international conference on industrial engineering and operations management* (pp. 793–800). IEOM.

Onu, P., & Mbohwa, C. (2018b). Sustainable oil exploitation versus renewable energy initiatives: A review of the case of Uganda. *Proceedings of the International Conference on Industrial Engineering and Operations Management*, 1008–1015 (Washington DC).

Onu, P., & Mbohwa, C. (2018c). Correlation between future energy systems and industrial revolutions. *Proceedings of the International Conference on Industrial Engineering and Operations Management*, 1953–1961 (Pretoria/Johannesburg).

Onu, P., & Mbohwa, C. (2019a). Sustainable production: New thinking for SMEs. *Journal of Physics: Conference Series, 1378*(2), Article 022072. https://doi.org/10.1088/1742-6596/1378/2/022072.

Onu, P., & Mbohwa, C. (2019b). Sustainable supply chain management: Impact of practice on manufacturing and industry development. *Journal of Physics: Conference Series, 1378*(2), Article 022073. https://doi.org/10.1088/1742-6596/1378/2/022073.

Onu, P., & Mbohwa, C. (2019c). *Advances in solar photovoltaic grid parity.* IEEE, Article 978-1-7281-5152-6.

Onu, P., & Mbohwa, C. (2019d). *Concentrated solar power technology and thermal energy storage : A brief overview of nascent sustainable designs.* IEEE, Article 978-1-7281-5152-6.

Onu, P., & Mbohwa, C. (2019e). Renewable energy technologies in brief. *International Journal of Scientific and Technology Research, 8*(10), 1283–1289.

Onu, P., & Mbohwa, C. (2019f). Industrial energy conservation initiative and prospect for sustainable manufacturing. *Procedia Manufacturing, 35*, 546–551. https://doi.org/10.1016/j.promfg.2019.05.077.

Onu, P., & Mbohwa, C. (2019g). *New future for sustainability and industrial development : Success in blockchain, internet of production, and cloud computing technology* (pp. 1–12). OSCM.

Onu, P., Abolarin, M. S., & Anafi, F. O. (2017). Assessment of effect of rice husk ash on burnt properties of badeggi clay. *International Journal of Advanced Research, 5*(5), 240–247. https://doi.org/10.21474/IJAR01/4103.

Papargyropoulou, E., Lozano, R., Steinberger, J. K., Wright, N., & Ujang, Z. Bin (2014). The food waste hierarchy as a framework for the management of food surplus and food waste. *Journal of Cleaner Production, 76*, 106–115. https://doi.org/10.1016/j.jclepro.2014.04.020.

Pattanaik, L., Pattnaik, F., Saxena, D. K., & Naik, S. N. (2019). Biofuels from agricultural wastes. In *Second and third generation of feedstocks. In Lopa* (pp. 103–142). Elsevier.

Pleissner, D., Sze, C., & Lin, K. (2013). *Valorisation of food waste in biotechnological processes* (pp. 1–6). Sustainable Chemical Processes.

Puyol, D., Batstone, D. J., Hülsen, T., Astals, S., Peces, M., & Krömer, J. O. (2017). Resource recovery from wastewater by biological technologies: Opportunities, challenges, and prospects. *Frontiers in Microbiology, 7*. https://doi.org/10.3389/fmicb.2016.02106.

Raji, I. O., & Onu, P. (2017). Untapped wealth potential in fruit for Uganda community. *International Journal Of Advanced Academic Research, 3*(February), 17–25. Retrieved from http://www.ijaar.org/articles/Volume3-Number1/Sciences-Technology-Engineering/ijaar-ste-v3n1-jan17-p7.pdf.

Ruane, J., & Sonnino, A. (2011). Agricultural biotechnologies in developing countries and their possible contribution to food security. *Journal of Biotechnology, 156*, 356–363. https://doi.org/10.1016/j.jbiotec.2011.06.013.

Ruane, J., Sonnino, A., & Agostini, A. (2010). Bioenergy and the potential contribution of agricultural biotechnologies in developing countries. *Biomass and Bioenergy, 34*, 1427–1439. https://doi.org/10.1016/j.biombioe.2010.04.011.

Salomone, R., Saija, G., Mondello, G., Giannetto, A., Fasulo, S., & Savastano, D. (2017). Environmental impact of food waste bioconversion by insects: Application of life cycle assessment to process using *Hermetia illucens*. *Journal of Cleaner Production, 140*, 890–905. https://doi.org/10.1016/j.jclepro.2016.06.154.

Sharma, B., Sarkar, A., Singh, P., & Singh, R. P. (2017). Agricultural utilization of biosolids: A review on potential effects on soil and plant grown. *Waste Management, 64*, 117–132. https://doi.org/10.1016/j.wasman.2017.03.002.

Skaggs, R. L., Coleman, A. M., Seiple, T. E., & Milbrandt, A. R. (2018). Waste-to-Energy biofuel production potential for selected feedstocks in the conterminous United States. *Renewable and Sustainable Energy Reviews, 82*, 2640–2651. https://doi.org/10.1016/j.rser.2017.09.107 (October 2017).

Soccol, C. R., Vandenberghe, L. P. de S., Medeiros, A. B. P., Karp, S. G., Buckeridge, M., Ramos, L. P., ... Torres, F. A. G. (2010). Bioethanol from lignocelluloses: Status and perspectives in Brazil. *Bioresource Technology*. https://doi.org/10.1016/j.biortech.2009.11.067.

UN-Water. (2016). The United nations world water development report 2016 on "water and jobs" main messages. In UN. In *Leaflet*. UNESCO.

UNFCC. (2015). Adoption of the paris agreement. In UN (Ed.), *Conference of the Parties on its twenty-first session*. United Nations. https://doi.org/FCCC/CP/2015/L.9.

Waldron, K. (2009). *Handbook of waste management and Co-product recovery in food processing. 1*. Woodhead. In Handbook of Waste Management and Co-Product Recovery in Food Processing.

Wang, H., Pu, Y., Ragauskas, A., & Yang, B. (2019). From lignin to valuable products—strategies, challenges, and prospects. *Bioresource Technology, 271*, 449–461. https://doi.org/10.1016/j.biortech.2018.09.072.

Wanyama, J., Banadda, N., Kiyimba, F., Okurut, S., Zziwa, A., Kabenge, I., ... Kiggundu, N. (2016). Profiling agricultural engineering technologies for mechanizing smallholder agriculture in Uganda. *Agricultural Engineering International: CIGR Journal*.

Woldeyohannes, A. D., Woldemichael, D. E., & Baheta, A. T. (2016). Sustainable renewable energy resources utilization in rural areas. *Renewable and Sustainable Energy Reviews, 66*, 1–9. https://doi.org/10.1016/j.rser.2016.07.013.

Xu, W., Fu, S., Yang, Z., Lu, J., & Guo, R. (2018). Improved methane production from corn straw by microaerobic pretreatment with a pure bacteria system. *Bioresource Technology, 259*, 18–23. https://doi.org/10.1016/j.biortech.2018.02.046.

Yusuf, A. A., Peter, O., Hassanb, A. S., Tunji., A. L., Oyagbola, I. A., Mustafad, M. M., & Yusuf, Danjuma A. (2019). Municipality solid waste management system for Mukono district. *Procedia Manufacturing, 35*, 613–622. https://doi.org/10.1016/j.promfg.2019.06.003.

Zheng, R., Su, S., Zhou, H., Yan, H., Ye, J., Zhao, Z., ... Fu, X. (2017). Antioxidant/antihyperglycemic activity of phenolics from sugarcane (*Saccharum officinarum* L.) bagasse and identification by UHPLC-HR-TOFMS. *Industrial Crops and Products*. https://doi.org/10.1016/j.indcrop.2017.03.012.

CHAPTER

Methodological approaches in agrowaste preparation and processes

3

3.1 Introduction

The pressing desire to find sustainable energy that will substitute fossils, amid the concern of energy potent farm crops that compete favorably with cash and food crops (first-generation biofuel resources), draws on the limitation to expand production of biofuels. New scientific interventions, through research and development, now explore varieties of the second-generation biofuel crops, termed as the lignocellulosic-based biomass (Lawal et al., 2019; Onu & Mbohwa, 2019a). The naturally available biodegradable agricultural resources are viable for sustainable biomethane production. However, the presence of lignin resists the degradation under anaerobic conditions of the lignocellulosic material and challenges its potential as a biofuel crop. Thus, the concern to enhance biomass conversion to remediate poor degradation of bioenergy raw material becomes a top priority toward achieving the near-future goal of a sustainable green ecosystem through cost-effective processes (Onu & Mbohwa, 2018a). That is to say, adequate preparation and means to convert the lignocellulosic biomass have been explored for its renewable energy characteristics. Pretreatment strategies and related processes that support enhanced biofuels production are presented as follows.

The exploitation of lignocellulosic biomass resources (LBR) and the supporting technologies that offer potential through which agricultural excesses can find usefulness exists for the following reasons. First, it is a renewable energy source that could be sustainably developed in the future. Next, its characteristics as a green resource support positive environmental features resulting in zero releases of carbonaceous gases and negligible amounts of sulfur to the atmosphere. Thirdly, it appears to have significant economic potential, inasmuch as the price of fossil fuel does not plunge in years to come (Achinas & Euverink, 2016; Onu & Mbohwa, 2019b). Biomass conversion to either gaseous or liquid fuels through the biological routes, or thermochemical process, has drawn research attention globally and aims to manage agricultural waste through careful waste disposal strategies and planning (Onu & Mbohwa, 2018b). The essence of producing sustainable and renewable energy from agricultural waste resources justifies the necessity for the preparation of the materials required in the process of increased production.

Agricultural Waste Diversity and Sustainability Issues. https://doi.org/10.1016/B978-0-323-85402-3.00009-7

The biomass characterization is necessary to determine its moisture, fixed carbon, volatile solids, and ash content, as part of the proximate composition analysis. Similarly, processes such as ultimate composition analysis will help to determine the potency of the feedstock, while the lignocellulose composition implies the presence of cellulose, hemicellulose, and lignin. The biochemical analysis is essential to determine the content level of carbohydrates, protein, lipids, and other elements. Pattanaik, Pattnaik, Saxena, and Naik (2019) have illustrated the aforementioned assertion, categorizing rice straw, wheat straw, corn stover, barley straw, and oat straw as crop residues, and determine their biochemical composition (wt.%) to have lipids contents of 5.9, 5.34, 0.7–1.3, 1.91, and 1.65 respectively. The report shows that the lignocellulosic composition (wt.%) of the feedstocks did not exceed 53% (cellulose), 34% (hemicellulose), and 22% (lignin) for the category of the crop residues tested.

Cellulose

Cellulose $(C_6H_{10}O_5)_n$ is a hexose sugar from agrowastes and biomasses, including wood. It is a linear polymer of glucose monomers (D-glucose) linked to β-(1,4)-glycosidic bonds and consists of a long chain of β-glucose monomers gathered into microfibril bundles (Haghighi Mood et al., 2013). The insoluble property of cellulose in water allows the process of hydrolysis to commence and involves the breakdown of polysaccharide to free sugar molecules known (saccharification), thus increasing the water content in the reaction.

Hemicellulose

Hemicellulose $(C_5H_8O_4)_n$ is a short, highly branched polymer of pentose sugars (D-xylose and L-arabinose) and hexose (D-glucose, D-mannose, and D-galactose) (Haghighi Mood et al., 2013). Hemicelluloses are associated with cellulose as a large source of carbon in plants and xyloglucans or xylans, depending on the types of plants. The presiding sources of hemicelluloses biomass are woody, softwoods, and hardwoods (Limayem & Ricke, 2012).

Lignin

Lignin $[C_9H_{10}O_3(OCH_3)_{0.9-1.7}]_n$ is an organic compound from three different monomers (coniferyl, synapyl alcohols, and p-coumaryl) linked together to form a matrix (Onu, Abolarin, & Anafi, 2017; Sánchez, 2009). The matrix consists of several functional groups: methoxyl, hydroxyl, and carbonyl, which shows a high polarity to the lignin macromolecule, which contributes to the nonfermentation of LCBm, leading to poor degradation of the agrowaste substrate and low quality of bioethanol production (Taherzadeh & Karimi, 2008).

The determination, therefore, of the retention potential of feedstock or in a case to optimize its yield calls for careful selection, handling, and pretreatment of the biomass. Feedstock characteristics that are supported by the thermochemical conversion require low moisture and ash content (García, Pizarro, Lavín, & Bueno, 2012). Syngas, biocoal, and biochar are all products of the thermochemical conversion process. It combines heating and chemical processes and minimizes environmental pollution. Crop residues produced during agricultural activities in the farms serve as direct wastes from an agricultural operation and leaves a high amount of agricultural wastes that require processing. The estimated crop residues produced annually are estimated above 2800 million tons, according to Zabed, Sahu, Boyce, and Faruq (2016). These residues are a readily available and considerably cheap form of agricultural wastes that have the potential for viable end product. The most sort-out agricultural residue that finds great use for bioethanol production is the corn stover, and the rice and wheat straws. These crops are abundant all through the year, leaving behind enormous waste during its processing; however, very little attention is given to its biofuel production potential.

The traditional African practice of handling most agricultural wastes is by dumping them in the farms, allowing to dry out, or burning after a short period, which leads to environmental pollution and causes breathing discomfort (Yusuf et al., 2019; Raji & Onu, 2017). Rice straw deposit has been identified as one of the most viable, abundant, and resourceful biomasses globally with an annual production capacity that exceeds 730 million tons, with the highest producers being Asia (Sarkar, Ghosh, Bannerjee, & Aikat, 2012). The value is not far-fetched from that of wheat straw and the corn stover, at 354.34 million tons and 128.02 million tons, respectively. Interestingly, sugarcane bagasse makes the list, contributing up to 180.73 million tons annually (Saini, Saini, & Tewari, 2015).

Remains from several food crops (vegetable, fruit peels, fruit pomace, sugarcane bagasse, seed cake, bones, and meat) after being processed become essential raw materials for the creation of diverse domestic and health products. Majorly, livestock wastes are classified into three: liquid manure, solid manure, and wastewater from the butchery section or other washing activities. These waste categories, when untreated, become hazardous and infect the air, water, and soil around. Particularly, the solid manure leaves the air in the surrounding unpleasant to organic life forms amid the effect of greenhouse gases (Onu & Mbohwa, 2019b), having great implications for sustainable human settlement.

3.2 Physical and chemical pretreatment of biomass feedstock

A lot of reason has been given to justify the reduction in particle size of feedstock as a preliminary exercise to allow increased surface area to facilitate subsequent activities such as microbial actions that facilitate fermentation and subsequently biofuels in production.

Physical pretreatment

The physical pretreatment process has been described, as seen in Fig. 3.1, which comprises mechanical milling, microwave pretreatment, and ultrasound processes. The mechanical pretreatment operation is essential for the remodification of the resources to improve the bioconversion potential through particle densification and distribution. It is a simple, nontoxic operation that follows lay-down procedures for size reduction (Alvira, Tomás-Pejó, Ballesteros, & Negro, 2010; Onu et al., 2017). In the physical−mechanical pretreatment process, there are no chemicals involved, and hence, it is a nonacidic process, and there are no inhibitors present (Zabed et al., 2016). The mechanical working process offers increased porosity and bulk density, which are vital criteria for the quality of the end product since it affects the performance of the conversion of the biomass (lignocellulose), exposing the cellulose and hemicellulose matrixes during hydrolysis (Zheng, Pan, & Zhang, 2009). The other two processes have a combined effect on the lignocellulose compound, where it dissociates the hemicellulose and lignin. As such, the microwave and ultrasound pretreatment operation is solely to assist in further drying of the biomass ready for digestion. Next, the degradation of the cellulosic structure is followed (Ultrasound), according to literature (Fernandes, Linhares, & Rodrigues, 2008; Çakmak, Tekeoğlu, Bozkir, Ergün, & Baysal, 2016). The process, however, has its flaws, as the event of milling and ultrasound or microwaving of the biomass feedstock may lead to cell rupture, resulting in the loss of time and energy.

Chemical pretreatment

The chemical pretreatment process adopts various techniques and use of different chemicals to prepare the feedstocks before being charged into a biodigester. The pretreatment process involves the use of alkalis and solvents that are organic in nature in a process called the organosolv chemical pretreatment (Mesa et al., 2011).

Acid pretreatment

The use of dilute acids finds application in this process as a pretreatment method and, thus, promotes degradation. Aqueous sulfuric acid has been tested to have

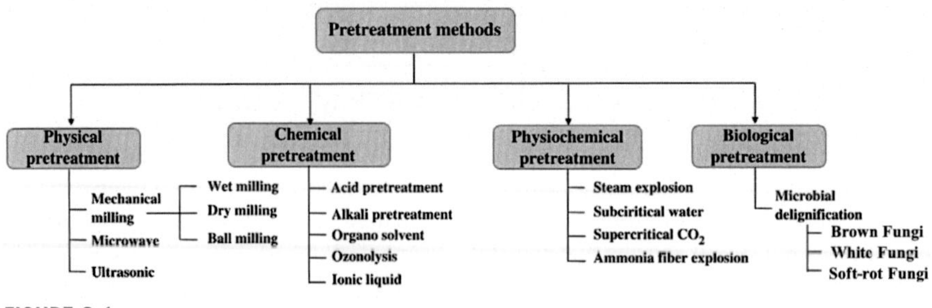

FIGURE 3.1

Pretreatment process of potential bioresources.

high efficiency in the separating process of cell wall components, resulting in hemicellulose hydrolysate and cellulignin (Dussán, Silva, Moraes, Arruda, & Felipe, 2014; Kumar & Sarkar, 2011). Other literature showcases hydrochloric, oxalic, and acetic acids, having an excellent prospect for efficient and effective biomass pretreatment (Chandel, da Silva, & Singh, 2011; Monlau, Latrille, Da Costa, Steyer, & Carrère, 2013; Pakarinen, Kaparaju, & Rintala, 2011). The process follows a standard protocol of no less than 15 minutes, with retention time, not exceeding 30 minutes, while operating below 180°C being adequate to subtle the cell and lead the degradation process. Regardless, this pretreatment activity has zero effect on the lignin matrix; however, this promotes the cellulosic and hemicellulose saccharification. More so, the pretreatment techniques are bound to lead the formation of inhibitors, dependent on the acidity of the "soaking" time.

Alkali pretreatment

This pretreatment technique is highly efficient, especially when using sodium hydroxide (NaOH), during the process of producing ethanol, with rapid conversion rate. The technological setup for the process is a low temperature and pressure arrangement, thereby leading to the negation of inhibitors in the system. Regardless, the effect due to low temperature also affects sugar degradation (Canilha et al., 2012). Other alkaline compounds such as potassium hydroxide (KOH), anhydrous ammonia (NH3), and calcium hydroxide Ca(OH)2) have also shown viable performances in the conversion of biomasses for biofuel production (Sebayang et al., 2016). This method is highly recommended for low lignin content feedstocks due to its capability to produce more than 80% glucose yield in its hydrolytic process (Bali, Meng, Deneff, Sun, & Ragauskas, 2015).

Ozonolysis pretreatment

Amid the progress of the chemical pretreatment processes, ozonolysis stands out as very promising, which has a high oxidative effect on lignocellulose biomasses during pretreatment, leading to greater yield in sugar production tenable for fermentation and later to obtain the desired biofuel (enzymatic hydrolysis) (Roncero, Torres, Colom, & Vidal, 2003). The oxidation capability of ozone allows for its application in diverse areas, including the pulp and paper industry. The ozone is soluble in water and unstable around compounds that have conjugated double bonds. Therefore, it directly attacks/oxidizes lignin due to the C=C bonds. Ozonolysis cleaves carbon–carbon bonds, which can occur at high temperatures or in catalytic beds, without having a direct impact on the ecosystem, where no harmful pollutants are released. In essence, the formation of short-chain carboxylic acids (inhibitors) is eminent. At the same time, the quality of the hydrolysate is dependent on the type of reactor, moisture condition inside the reactor, particle size, acidity level, reaction time, ozone/airflow, and ozone concentration (Travaini, Martín-Juárez, Lorenzo-Hernando, & Bolado-Rodríguez, 2016).

Ionic liquids

These are solvents of salts made from organic and inorganic compositions of anions and cations. The ionic liquid is mainly categorized into four: N-alkylpyridinium, N-alkyl-isoquinolinium, 1-alkyl-3-methylimidazolium, and the quaternary ammonium (Liu, Wang, Stiles, & Guo, 2012). The desired properties of ionic liquids are high thermochemical stability and low vapor pressure property. Nonetheless, other physical factors affect the choice of ionic liquid in pretreatment. These include corrosiveness, toxic nature, affordability, particular behavior in water, and, if biodegradable, to mention few (Mäki-Arvela, Anugwom, Virtanen, Sjöholm, & Mikkola, 2010). Challenges of adopting this technic range from a list of issues, which include but not limited to, (1) the dissolution of cellulose by the ionic liquid, (2) possible cell structure modification, and (3) interference of residual ionic liquid with the hydrolytic process, reducing the performance of sugar production and, hence affecting final biofuel yield (Liu et al., 2012; Sathitsuksanoh, Zhu, & Zhang, 2012).

Combined, physical and chemical pretreatment

This process is commonly accepted and termed as the physicochemical pretreatment process. Thus, it combines both the physical and chemical approaches. The physicochemical techniques include steam explosion, subcritical water (liquid hot water treatment), supercritical CO_2 (carbon dioxide explosion), and the ammonia fiber explosion.

The "steam explosion" is one of the most popular, efficient, and cost-effective pretreatment techniques for feedstocks and lignocellulosic biomasses. High saturated steam, which does not exceed 250°C and 50atm, is used to bombard the raw material, which causes separation of the individual fiber and swelling of the legionellae to disrupt their cell structure to promote the subsequent hydrolysis process (Rodriguez, Alaswad, Benyounis, & Olabi, 2017). An early illustration of this technique was demonstrated by Cara et al. (2008), where improvement in enzymatic digestibility was observed, and the pretreated lignocellulosic-based materials were discovered to reserve a high amount of hemicellulose sugar.

The "carbon dioxide explosion" is also a technique that disrupts the cellulosic structure of bioenergy feedstocks for the purpose of increasing the accessibility and substrate surface area during the enzymatic hydrolysis process through the use of pressurized carbon dioxide gas (Morais, Da Costa Lopes, & Bogel-Łukasik, 2015). This process is maintained at a lower temperature compared with the steam explosion process and saves cost on energy consumption and no adverse effect on the decomposition of monosaccharides. More so, the utilization of carbon dioxide gas is essential due to its low cost and no toxicity property. It is soluble in water and further facilitates the hydrolysis of hemicellulose (Zheng, Zhao, Xu, & Li, 2014).

The "liquid hot water technique" is a pretreatment strategy, which supports the dissociation of lignocellulose biomasses by using hot water (160°C to 240°C). The procedure creates the appearance of a hydronium ion that facilitates the release of acetic acid from hemicelluloses, thereby improving its removal. Regardless, the process promotes the presence of inhibitor, which lowers enzymatic activities and decreases performance based on the conversion of cellulose to glucose (Imman, Arnthong, Burapatana, Laosiripojana, & Champreda, 2013; Patil et al., 2013). Finally, the ammonia fiber explosion (AFEX) process adopts the use of ammonia to suppress the crystallinity of lignocellulosic biomasses and hence increases cell disability and promotes enzymatic hydrolysis (Maurya, Singla, & Negi, 2015).

3.3 Biological pretreatment of agrowaste

Pretreatment is an essential step that combines different techniques for processing lignocellulosic biomass to ensure high yield production of biofuel. The processes of preparation of the biomass help in the delignification and decomposition of the lignocellulose composites and also ensure even substrate size and requirements based on other characteristics features of the substrates to be met (Sarkar et al., 2012). Therefore, microbial presence is needed to delignify or assist in the decomposition process of hemicellulose, which is an essential activity for the success of agrowaste conversion to biofuels, and form part of the biological pretreatment activity. Efficient decomposition of the lignocellulose is desired for improved hydrolysis and is made possible by the presence of fungal consortium for the removal of lignin from agricultural feedstocks (Maurya et al., 2015; Song, Yu, Ma, & Zhang, 2013). Rice husks and corn stalks have been observed to have their lignin content reduced by microorganisms: *Phanerochaete chrysosporium* and *Irpex lacteus* in an efficient high yield sugar reducing process (Taha et al., 2015). The process where the hemicellulose composition of the lignocellulosic material is delignified and decomposition takes place enhances cellulose production in the presence of the other vital monomers, which aid in the anaerobic digestion process. The different techniques used in biological pretreatment for sustainable biofuel/biogas production have utilized bacteria, fungi, and microbial consortia, among others, to improve biofuel yield in an anaerobic digestion process. Different concepts in biological pretreatment strategies are presented in this section, with emphasis on how the processes can contribute to the effective anaerobic condition. Thus, pretreatment is conducted, bearing in mind the major aims, to deconstruct the lignin in the lignocellulosic matrix, reduce the crystallinity of cellulose and enlarge the biomass surface area.

Bacteria pretreatment

The bacteria pretreatment approach is known to be one of the fastest processes for biofuel production, giving a rapid rate of metabolism. Nonetheless, the degradation of lignin is not quite as effective when compared with the other techniques of using fungi and microbial consortia. Some categories of bacteria have been found to produce certain essential enzymes (aerobic and hydrolytic bacteria), which may facilitate the entire lignocellulose conversion and lignin-degrading process (Lü, Ji, Shao, & He, 2013; Xu, Fu, Yang, Lu, & Guo, 2018). Biomass pretreatment using bacterial is fast-gaining application to increase yield in methane production from anaerobic digestion. This has been experimented using hydrolytic bacteria on biomass slurry, which yielded very high methane production (Fu, Wang, Shi, & Guo, 2016).

Fungi pretreatment

Fungi are known to have a high enzymatic potential for biological pretreatment and conversion. Classified differently (brown rot, white rot, and soft rot), fungi have been known with very high degrading capability in the treatment of lignocellulose biomasses, e.g., the Basidiomycota, consisting of the white rot and brown rot fungi. While the white-rot fungi are known to have effective lignin degradation potential, they also have a high capability to disintegrate cellulose and promote enzyme activities. Regardless, a recent study shows that some fungi increase the rate of cellulose decomposition (Rodriguez et al., 2017). For example, "Pleurotus ostreatus," selectively degrade lignin and reduced the disintegration/loss of cellulose and hemicellulose, thereby, proven to be an efficient biological delignifying fungi, when considering the function of loss of cellulose against lignin degradation (Kamcharoen, Champreda, Eurwilaichitr, & Boonsawang, 2014).

Microbial consortia pretreatment

This process involves the mixture of microorganisms that work cooperatively, as per the pretreatment of biomass material (Sindhu, Binod, & Pandey, 2016). The need to increase metabolic activities in the substrates applies the synergic interaction of two or more bacteria and fungi to contribute to the rapid degradation in the biofuel production process over a short period. The technology for microbial consortia is favorable to sustain hydrolysis efficient and increased production under controlled pH, while the substrates are optimally utilized. However, the limitation of this pretreatment approach requires a high level of technicality and expertise to maintain the operating conditions of the system and the microorganism interactions. The process has extensively been used for the production of methane biogas and bioethanol production. More recent studies have investigated a consortium system of multiple bacterial, fungi, and actinomycetes strain to access lignin degradation (Fang et al., 2018). Their result shows a tremendously positive effect on tree trimmings. The process is sustainable for biomass pretreatment to significantly assist biodegradation through changing their complex crystalline structure and molecular forms.

Microaerobic pretreatment

The condition of having a low concentration of air in an anaerobic process has been investigated at different oxygen levels to access the microbial activity for the assimilation of hydrolysis and found to be effective in agrowaste water and sewage treatment (Montalvo, Vielma, Borja, Huiliñir, & Guerrero, 2018). The microaeration processes activate microbial action during hydrolysis and facilitate the digestion of different forms of agricultural wastes and energy crops.

Ensilage process

This is a process that helps to store and concurrently treat biomass. Solid-state fermentation is initiated during this process, promoting conditions that aid in the anaerobic digestion process. This strategy ensures the availability of a stabilized lignocellulosic matrix with low lignin composition for anaerobic conversation and biofuel production. An investigation into a wet ensilage process where biomasses are stored and concurrently treated with chemicals and fungal strains shows increased glucose and xylose production (Cui, Shi, Wan, & Li, 2012), and in recent times, researchers have reported a positive outcome of their investigations with different biomasses (Prapinagsorn, Sittijunda, & Reungsang, 2017; Wagner, Janetschek, & Illmer, 2018). Regardless of the ensiling effects and the chemical influences, the success of the process is greatly dependent on the structure and composition of the biomasses (mostly effective for sugar-based biomasses like the sweet corn stoves, sorghum stalks, etc.).

Enzymes process

This process performs closely to the fungal pretreatment for lignin degradation. The enzymatic activity includes the use of ligninolytic or hydrolytic enzymes for the delignification of lignocellulose biomasses. It is a cost-effective process from the perspective of operability and productivity, according to Asgher, Ahmad, and Iqbal (2013). Recent investigation has researched the utilization of enzymes (enzyme cocktail), exploring the single, and mixture of varying bacteria and fungi strains to obtain better results with opportunities for the futuristic biomass treatment (Agnolucci et al., 2019; Hui-Min et al., 2018).

Biochemical processes and biomass conversion routes

The formation of some toxic inhibitors appears during the conversion process of lignocellulose biomasses, like the phenolic and furan derivatives which ought to be removed through the process detoxification (Fig. 3.2), occurring after chemical, physical, and physicochemical pretreatments have been applied (Ibrahim, Ramli, Kamal Bahrin, & Abd-Aziz, 2017). As such, these inhibitors have been identified to be detrimental for the success of hydrolysis, or in extreme cases, whereof acidic medium, where corrosion of equipment begins to propagate all through the system. The degradation of lignin is most effective through the biological pretreatment process compared with other processes. The pretreatment process does not require additional steps for residual streams and presents itself as a viable alternative approach (Carrillo-Nieves et al., 2019; Kaida et al., 2009).

The most popular pretreatment process carried out to ensure hydrolysis in lignocellulose biomass is the enzymatic conversion. However, the cost implication of

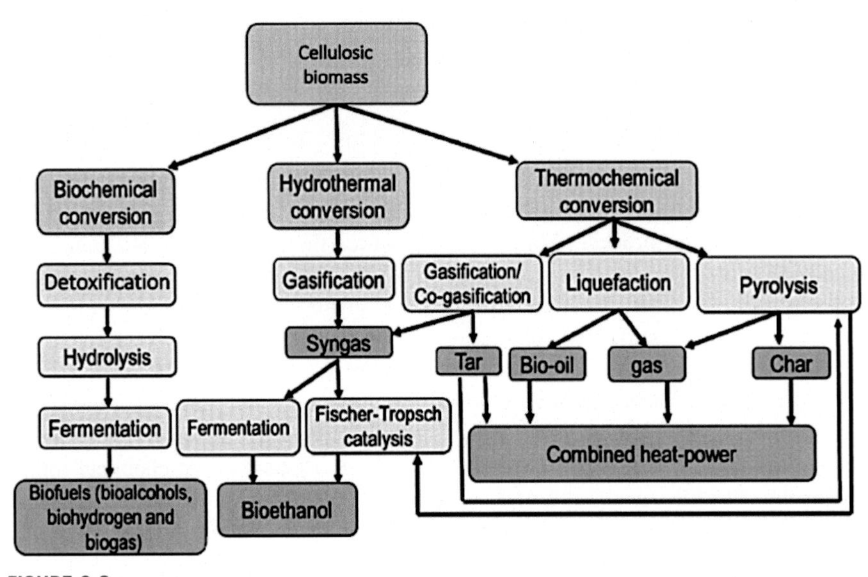

FIGURE 3.2

Showing stages of cellulosic biomass conversion.

acquiring the required enzymes (cellulase and hemicellulases) is considerably high, which is above $25 USD per cubic meter of producing ethanol (Chovau, Degrauwe, & Van Der Bruggen, 2013). Thus, newer alternatives, which are cost-effective and technologically up-to-date, may be explored for the production of high quality, energy-efficient, and sustainable biofuel. While it has been noticed that the processes of pretreating feedstocks for biofuel production, such as the enzymatic hydrolysis and fermentation are conducted differently, some researchers have succeeded in integrating enzymatic saccharification and fermentation in a single operation (Jahnavi, Prashanthi, Sravanthi, & Rao, 2017; Paulova, Patakova, Branska, Rychtera, & Melzoch, 2015).

Detoxification

The detoxification process of agricultural waste resources for biofuel production is purposeful for high yield of bioethanol and biobutanol. Detoxification involves the process of separating elements, which serve to hinder enzymes and microbial activities in the lignocellulosic raw materials to speed fermentation activities. Enzymatic and microorganism inhibitors are formed during the degradation of the lignin during the biological conversion process, hindering sustainable and efficient biofuel production. The causes and effects of the inhibitors do not support the long-term efforts of biofuel development from agrowastes material and have resulted in different techniques for their removal. A chemical treatment approach utilizes $Ca(OH)_2$, NaOH, and NH_4OH, as reducing agents. Other procedures such as the liquid–liquid,

liquid–solid extractions, and the use of laccase and peroxidase as enzyme treatment have proven to be a valuable approach to reducing the inhibitor activity (Jönsson, Alriksson, & Nilvebrant, 2013). In the case of the cost consideration, and ease of operation, other techniques to tackle the inhibitor challenge are through the identification and appropriate selection of valuable feedstocks. The aim is to access biomass with less tendency to generate high inhibitor or the need to utilize less slowly biodegradable or nonbiodegradable (recalcitrant biomass). Moreover, scientifically modified microbes have the potential to overcome the limitation of detoxification and, hence, the necessity to improve the yield of biofuel production.

Hydrolysis

The hydrolysis process utilizes acid/enzymes for the treatment of lignocellulosic biomass. The pretreatment process is such as to degrade the lignocellulosic composition into a monomeric unit. The acidic hydrolysis is carried out in two ways: under high temperature and pressure in the presence of a low concentration acid solution, and under the low temperature condition of high acidity. While hemicellulose is degraded under the dilute acid condition and, thus, undergoes pretreatment, the cellulose, in concentrated acid hydrolysis, is generally targeted at the degradation of cellulose and hemicellulose. Research has been conducted on hydrolysis in high acidic concentration with an excellent level of conversion of cellulose and hemicellulose to sugar (saccharification process) compared with dilute acid hydrolysis where the latter is less cost-effective. Nonetheless, the processes are faced with corrosion challenges of equipment in an acidic environment (Balat, 2011). Thus, the appreciation of the enzymatic hydrolysis process, where artificially introduced enzymes (cellulases, hemicellulases, and lignanases) influence the process operation and facilitate the degradation of lignocellulosic compounds, is possible.

The success of the enzymatic hydrolysis process is dependent: (1) the nature of the enzyme and (2) the substrate. The enzyme selection is based on several characteristics: type, source, and efficiency of the enzyme, whereas the substrate selection requirement considers the feedstock/particle size, porosity, and cellulose potential.

Fermentation

The hydrolysis stage, where sugar is released (saccharification process), ushers in the fermentation process. The activity, therefore, will lead to the conversion of other valuable by-products and is performed through various approaches (Albergaria & Arneborg, 2016; Stanbury, Whitaker, & Hall, 2017). The strategies involved in integrating the saccharification stage with fermentation follow four important approaches:

i. Sequential hydrolysis and fermentation: In this process, the hydrolysis of the substrate is conducted differently from the fermentation process, both at different stages, one before the other, and combined for distillation to produce ethanol. The end product in this process is mostly inhibited or contaminated and therefore results in a lesser yield of the ethanol/butanol (Demichelis et al., 2017; Song et al., 2019).

ii. Simultaneous saccharification and fermentation. In this process, there is simultaneous fermentation of glucose and the hydrolysis of cellulose in the same reactor. It is a cost-effective (enzymatic loading rate) and high-yield process, without the limitations with regard to inhibition and contamination as the sequential hydrolysis and fermentation. However, the challenges that affect the simultaneous saccharification and fermentation strategy arise in the case of a possible reason to simultaneously optimize the process conditions for hydrolysis (enzyme) and fermentation to take place (Chen & Fu, 2016).
iii. Simultaneous saccharification and cofermentation: The process adapts selected microorganisms (e.g., *coculture* of Zymomonas Mobilis and Candida *Shehata*), which is suitable and compactable with the system operating temperature and pH condition. As such, cofermentation of the saccharification compounds is achieved in the process (Robak & Balcerek, 2018).
iv. Direct microbial conversion: The process is a cost-effective way of managing enzymes' requirements for the production of bioethanol or butanol. Enzymes activities required during this operation, from hydrolysis to fermentation, are met by bacterial and fungal presence. The conversion process is one of the most economically feasible options among the earlier mentioned strategies. However, the process offers a much lower ethanol yield with a longer fermentation duration (Chen & Fu, 2016; Himmel, 2015; Zhao, Liu, Deng, & Zhu, 2017).

Improving operations in biological pretreatment

Innovative research and development in microorganisms interaction and performance for biofuel production are gaining prominence globally (Onu & Mbohwa, 2019c). The challenges that limit effective biological pretreatment of substrates and agrowaste materials underpin on the characteristics of the biomass, process condition, and the causative agents, which relates to size, moisture content, time, temperature, porosity, and type of enzymes. Reduction in the sizes of substrates will mean an increased surface area but causes congestion and leaves no room for efficient microbial activities. Consequently, an increase in the particle size limits penetration and leads to reduced efficiency of the microorganisms during pretreatment. More so, with regard to time constraints, research has shown greater yield in the hydrolysis process of lignocellulosic compounds. The longer the periods, the more the yield of sugar; however, this may vary depending on the pretreatment approach as fungi take a longer period (Cui et al., 2012). Furthermore, the tolerance of elevated temperature by any category of microorganism is an essential factor to consider to achieve microbial growth. While some process generates sufficient heat to support metabolism in the system, another process may require a slight increase in temperature of the medium to facilitate the digestive activity.

The major aspect that has been emphasized upon in this chapter targets the progressiveness of the hydrolytic and fermentation processes to reduce the lignin effect on the conversion of energy crops to gain biofuels. Biological pretreatment and improvement in its processes now focus on sustainable and robust technologies that are highly efficient, cost-effective, and easily operable (Onu & Mbohwa,

2019b). Also, the route to the optimal approach of pretreatment should form the scope of future research. Mixed pretreatment strategies are yet to be explored as there are still very few conceptualizations on combining either the physical, chemical, and physiochemical pretreatment with the biological pretreatment (Mustafa, Poulsen, Xia, & Sheng, 2017), thus requiring improvement. The potential of this suggestion is to explore rapid lignin degeneration for more cellulose retention and reduce pretreatment operation time.

3.4 Conclusions and prospects

Technological innovation is on the rise to effectively manage resources in agricultural development, leading to increased investment in biodiversity. The integrity of agrowastes development for clean energy with regard to effective pretreatment operations has been reviewed. Developing countries are on the receiving end with issues concerning poverty, food shortages, and poor agricultural waste management practices, mostly because of the financial implications to install and maintain massive biorefineries that are essential for the conversion of the agrowastes, and production of biofuels cannot be met. Feedstock availability, in comparison with capital investments and operational cost of biofuel production plants, needs to be evaluated to forecast the return on investment of the agrowaste beneficiation, including the most preferred method of pretreatment. Regional factors, infrastructural compatibility, road network, and market demand, among others, also form part of the pivotal element for consideration before arriving at the big picture. More studies are required to manage the technological deployment of biofuel production in Africa by conducting both fundamental and applied types of research.

References

Achinas, S., & Euverink, G. J. W. (2016). Consolidated briefing of biochemical ethanol production from lignocellulosic biomass. *Electronic Journal of Biotechnology*. https://doi.org/10.1016/j.ejbt.2016.07.006.

Agnolucci, M., Avio, L., Pepe, A., Turrini, A., Cristani, C., Bonini, P., … Giovannetti, M. (2019). Bacteria associated with a commercial mycorrhizal inoculum: Community composition and multifunctional activity as assessed by illumina sequencing and culture-dependent tools. *Frontiers in Plant Science*. https://doi.org/10.3389/fpls.2018.01956.

Albergaria, H., & Arneborg, N. (2016). Dominance of *Saccharomyces cerevisiae* in alcoholic fermentation processes: Role of physiological fitness and microbial interactions. *Applied Microbiology and Biotechnology*. https://doi.org/10.1007/s00253-015-7255-0.

Alvira, P., Tomás-Pejó, E., Ballesteros, M., & Negro, M. J. (2010). Pretreatment technologies for an efficient bioethanol production process based on enzymatic hydrolysis: A review. *Bioresource Technology*. https://doi.org/10.1016/j.biortech.2009.11.093.

Asgher, M., Ahmad, Z., & Iqbal, H. M. N. (2013). Alkali and enzymatic delignification of sugarcane bagasse to expose cellulose polymers for saccharification and bio-ethanol production. *Industrial Crops and Products*. https://doi.org/10.1016/j.indcrop.2012.10.005.

Balat, M. (2011). Production of bioethanol from lignocellulosic materials via the biochemical pathway: A review. *Energy Conversion and Management*. https://doi.org/10.1016/j.enconman.2010.08.013.

Bali, G., Meng, X., Deneff, J. I., Sun, Q., & Ragauskas, A. J. (2015). The effect of alkaline pretreatment methods on cellulose structure and accessibility. *ChemSusChem*. https://doi.org/10.1002/cssc.201402752.

Çakmak, R.Ş., Tekeoğlu, O., Bozkir, H., Ergün, A. R., & Baysal, T. (2016). Effects of electrical and sonication pretreatments on the drying rate and quality of mushrooms. *Lebensmittel-Wissenschaft und -Technologie- Food Science and Technology*. https://doi.org/10.1016/j.lwt.2016.01.032.

Canilha, L., Chandel, A. K., Suzane Dos Santos Milessi, T., Antunes, F. A. F., Luiz Da Costa Freitas, W., Das Graças Almeida Felipe, M., & Da Silva, S. S. (2012). Bioconversion of sugarcane biomass into ethanol: An overview about composition, pretreatment methods, detoxification of hydrolysates, enzymatic saccharification, and ethanol fermentation. *Journal of Biomedicine and Biotechnology*. https://doi.org/10.1155/2012/989572.

Cara, C., Ruiz, E., Ballesteros, M., Manzanares, P., Negro, M. J., & Castro, E. (2008). Production of fuel ethanol from steam-explosion pretreated olive tree pruning. *Fuel*. https://doi.org/10.1016/j.fuel.2007.05.008.

Carrillo-Nieves, D., Rostro Alanís, M. J., de la Cruz Quiroz, R., Ruiz, H. A., Iqbal, H. M. N., & Parra-Saldívar, R. (2019). Current status and future trends of bioethanol production from agro-industrial wastes in Mexico. *Renewable and Sustainable Energy Reviews*. https://doi.org/10.1016/j.rser.2018.11.031.

Chandel, K. A., da Silva, S. S., & Singh, V. O. (2011). Detoxification of lignocellulosic hydrolysates for improved bioethanol production. In *Biofuel production-recent developments and prospects*. https://doi.org/10.5772/16454.

Chen, H., & Fu, X. (2016). Industrial technologies for bioethanol production from lignocellulosic biomass. *Renewable and Sustainable Energy Reviews*. https://doi.org/10.1016/j.rser.2015.12.069.

Chovau, S., Degrauwe, D., & Van Der Bruggen, B. (2013). Critical analysis of techno-economic estimates for the production cost of lignocellulosic bio-ethanol. *Renewable and Sustainable Energy Reviews*. https://doi.org/10.1016/j.rser.2013.05.064.

Cui, Z., Shi, J., Wan, C., & Li, Y. (2012). Comparison of alkaline- and fungi-assisted wet-storage of corn stover. *Bioresource Technology*. https://doi.org/10.1016/j.biortech.2012.01.037.

Demichelis, F., Pleissner, D., Fiore, S., Mariano, S., Navarro Gutiérrez, I. M., Schneider, R., & Venus, J. (2017). Investigation of food waste valorization through sequential lactic acid fermentative production and anaerobic digestion of fermentation residues. *Bioresource Technology*. https://doi.org/10.1016/j.biortech.2017.05.174.

Dussán, K. J., Silva, D. D. V., Moraes, E. J. C., Arruda, P. V., & Felipe, M. G. A. (2014). Dilute-acid hydrolysis of cellulose to glucose from sugarcane bagasse. *Chemical Engineering Transactions*. https://doi.org/10.3303/CET1438073.

Fang, X., Li, Q., Lin, Y., Lin, X., Dai, Y., Guo, Z., & Pan, D. (2018). Screening of a microbial consortium for selective degradation of lignin from tree trimmings. *Bioresource Technology*. https://doi.org/10.1016/j.biortech.2018.01.058.

Fernandes, F. A. N., Linhares, F. E., & Rodrigues, S. (2008). Ultrasound as pre-treatment for drying of pineapple. *Ultrasonics Sonochemistry*. https://doi.org/10.1016/j.ultsonch.2008.03.009.

Fu, S. F., Wang, F., Shi, X. S., & Guo, R. B. (2016). Impacts of microaeration on the anaerobic digestion of corn straw and the microbial community structure. *Chemical Engineering Journal*. https://doi.org/10.1016/j.cej.2015.11.070.

García, R., Pizarro, C., Lavín, A. G., & Bueno, J. L. (2012). Characterization of Spanish biomass wastes for energy use. *Bioresource Technology*. https://doi.org/10.1016/j.biortech.2011.10.004.

Haghighi Mood, S., Hossein Golfeshan, A., Tabatabaei, M., Salehi Jouzani, G., Najafi, G. H., Gholami, M., & Ardjmand, M. (2013). Lignocellulosic biomass to bioethanol, a comprehensive review with a focus on pretreatment. *Renewable and Sustainable Energy Reviews*. https://doi.org/10.1016/j.rser.2013.06.033.

Himmel, M. E. (2015). *Direct microbial conversion of biomass to advanced biofuels*. https://doi.org/10.1016/c2011-0-09652-7.

Hui-Min, Y., Yan-Ping, W., Lin, L. Y., Shamsi, B. H., Bo, H., & Xu-Chun, M. (2018). Analysis of distribution and antibiotic resistance of pathogens isolated from the paediatric population in Shenmu hospital from 2011–2015. *Journal of International Medical Research*. https://doi.org/10.1177/0300060517716343.

Ibrahim, M. F., Ramli, N., Kamal Bahrin, E., & Abd-Aziz, S. (2017). Cellulosic biobutanol by Clostridia: Challenges and improvements. *Renewable and Sustainable Energy Reviews*. https://doi.org/10.1016/j.rser.2017.05.184.

Imman, S., Arnthong, J., Burapatana, V., Laosiripojana, N., & Champreda, V. (2013). Autohydrolysis of tropical agricultural residues by compressed liquid hot water pretreatment. *Applied Biochemistry and Biotechnology*. https://doi.org/10.1007/s12010-013-0320-1.

Jahnavi, G., Prashanthi, G. S., Sravanthi, K., & Rao, L. V. (2017). Status of availability of lignocellulosic feed stocks in India: Biotechnological strategies involved in the production of bioethanol. *Renewable and Sustainable Energy Reviews*. https://doi.org/10.1016/j.rser.2017.02.018.

Jönsson, L. J., Alriksson, B., & Nilvebrant, N. O. (2013). Bioconversion of lignocellulose: Inhibitors and detoxification. *Biotechnology for Biofuels*. https://doi.org/10.1186/1754-6834-6-16.

Kaida, R., Kaku, T., Baba, K., Oyadomari, M., Watanabe, T., Hartati, S., … Hayashi, T. (2009). Enzymatic saccharification and ethanol production of Acacia mangium and Paraserianthes falcataria wood, and Elaeis guineensis trunk. *Journal of Wood Science*. https://doi.org/10.1007/s10086-009-1038-0.

Kamcharoen, A., Champreda, V., Eurwilaichitr, L., & Boonsawang, P. (2014). Screening and optimization of parameters affecting fungal pretreatment of oil palm empty fruit bunch (EFB) by experimental design. *International Journal of Energy and Environmental Engineering*. https://doi.org/10.1007/s40095-014-0136-y.

Kumar, A., & Sarkar, S. (2011). *Biofuels – alternative feedstocks and conversion processes*.

Lawal, A. Q. T., Ninsiima, E., Odebiyib, O. S., Hassan, A. S., Oyagbola, I. A., Onu, P., & Yusuf, A. D. (2019). Effect of unburnt rice husk on the properties of concrete. *Procedia Manufacturing, 35*, 635–640. https://doi.org/10.1016/j.promfg.2019.06.006.

Limayem, A., & Ricke, S. C. (2012). Lignocellulosic biomass for bioethanol production: Current perspectives, potential issues and future prospects. *Progress in Energy and Combustion Science*. https://doi.org/10.1016/j.pecs.2012.03.002.

Liu, C. Z., Wang, F., Stiles, A. R., & Guo, C. (2012). Ionic liquids for biofuel production: Opportunities and challenges. *Applied Energy*. https://doi.org/10.1016/j.apenergy.2011.11.031.

Lü, F., Ji, J., Shao, L., & He, P. (2013). Bacterial bioaugmentation for improving methane and hydrogen production from microalgae. *Biotechnology for Biofuels*. https://doi.org/10.1186/1754-6834-6-92.

Mäki-Arvela, P., Anugwom, I., Virtanen, P., Sjöholm, R., & Mikkola, J. P. (2010). Dissolution of lignocellulosic materials and its constituents using ionic liquids-a review. *Industrial Crops and Products*. https://doi.org/10.1016/j.indcrop.2010.04.005.

Maurya, D. P., Singla, A., & Negi, S. (2015). An overview of key pretreatment processes for biological conversion of lignocellulosic biomass to bioethanol. *3 Biotech*. https://doi.org/10.1007/s13205-015-0279-4.

Mesa, L., González, E., Cara, C., González, M., Castro, E., & Mussatto, S. I. (2011). The effect of organosolv pretreatment variables on enzymatic hydrolysis of sugarcane bagasse. *Chemical Engineering Journal*. https://doi.org/10.1016/j.cej.2011.02.003.

Monlau, F., Latrille, E., Da Costa, A. C., Steyer, J. P., & Carrère, H. (2013). Enhancement of methane production from sunflower oil cakes by dilute acid pretreatment. *Applied Energy*. https://doi.org/10.1016/j.apenergy.2012.06.042.

Montalvo, S., Vielma, S., Borja, R., Huiliñir, C., & Guerrero, L. (2018). Increase in biogas production in anaerobic sludge digestion by combining aerobic hydrolysis and addition of metallic wastes. *Renewable Energy*. https://doi.org/10.1016/j.renene.2018.02.004.

Morais, A. R. C., Da Costa Lopes, A. M., & Bogel-Łukasik, R. (2015). Carbon dioxide in biomass processing: Contributions to the green biorefinery concept. *Chemical Reviews*. https://doi.org/10.1021/cr500330z.

Mustafa, A. M., Poulsen, T. G., Xia, Y., & Sheng, K. (2017). Combinations of fungal and milling pretreatments for enhancing rice straw biogas production during solid-state anaerobic digestion. *Bioresource Technology*. https://doi.org/10.1016/j.biortech.2016.11.028.

Onu, P., Abolarin, M. S., & Anafi, F. O. (2017). Assessment of effect of rice husk ash on burnt properties of badeggi clay. *International Journal of Advanced Research, 5*(5), 240–247. https://doi.org/10.21474/IJAR01/4103.

Onu, P., & Mbohwa, C. (2018a). Future energy systems and sustainable emission control: Africa in perspective. *Proceedings of the International Conference on Industrial Engineering and Operations Management*.

Onu, P., & Mbohwa, C. (2018b). Sustainable oil exploitation versus renewable energy Initiatives : A review of the case of Uganda. *Proceedings of the International Conference on Industrial Engineering and Operations Management*, 1008–1015 (Washington DC).

Onu, P., & Mbohwa, C. (2019a). Sustainable production: New thinking for SMEs. *Journal of Physics: Conference Series, 1378*(2). https://doi.org/10.1088/1742-6596/1378/2/022072.

Onu, P., & Mbohwa, C. (2019b). Renewable energy technologies in brief. *International Journal of Scientific & Technology Research, 8*(10), 1283–1289.

Onu, P., & Mbohwa, C. (2019c). Industrial energy conservation initiative and prospect for sustainable manufacturing. *Procedia Manufacturing, 35*, 546–551. https://doi.org/10.1016/j.promfg.2019.05.077.

Pakarinen, O. M., Kaparaju, P. L. N., & Rintala, J. A. (2011). Hydrogen and methane yields of untreated, water-extracted and acid (HCl) treated maize in one- and two-stage batch assays. *International Journal of Hydrogen Energy*. https://doi.org/10.1016/j.ijhydene.2011.08.028.

Patil, P., Reddy, H., Muppaneni, T., Ponnusamy, S., Sun, Y., Dailey, P., … Deng, S. (2013). Optimization of microwave-enhanced methanolysis of algal biomass to biodiesel under temperature controlled conditions. *Bioresource Technology*. https://doi.org/10.1016/j.biortech.2013.03.118.

Pattanaik, L., Pattnaik, F., Saxena, D. K., & Naik, S. N. (2019). Biofuels from agricultural wastes. In *Second and third generation of feedstocks*. https://doi.org/10.1016/b978-0-12-815162-4.00005-7.

Paulova, L., Patakova, P., Branska, B., Rychtera, M., & Melzoch, K. (2015). Lignocellulosic ethanol: Technology design and its impact on process efficiency. *Biotechnology Advances*. https://doi.org/10.1016/j.biotechadv.2014.12.002.

Prapinagsorn, W., Sittijunda, S., & Reungsang, A. (2017). Co-digestion of napier grass and its silage with cow dung for methane production. *Energies*. https://doi.org/10.3390/en10101654.

Raji, I. O., & Onu, P. (2017). Untapped wealth potential in fruit for Uganda community. *International Journal of Advanced Academic Research, 3*(February), 17–25. Retrieved from http://www.ijaar.org/articles/Volume3-Number1/Sciences-Technology-Engineering/ijaar-ste-v3n1-jan17-p7.pdf.

Robak, K., & Balcerek, M. (2018). Review of second generation bioethanol production from residual biomass. *Food Technology and Biotechnology*. https://doi.org/10.17113/ftb.56.02.18.5428.

Rodriguez, C., Alaswad, A., Benyounis, K. Y., & Olabi, A. G. (2017). Pretreatment techniques used in biogas production from grass. *Renewable and Sustainable Energy Reviews*. https://doi.org/10.1016/j.rser.2016.02.022.

Roncero, M. B., Torres, A. L., Colom, J. F., & Vidal, T. (2003). TCF bleaching of wheat straw pulp using ozone and xylanase. Part A: Paper quality assessment. *Bioresource Technology*. https://doi.org/10.1016/S0960-8524(02)00224-9.

Saini, J. K., Saini, R., & Tewari, L. (2015). Lignocellulosic agriculture wastes as biomass feedstocks for second-generation bioethanol production: Concepts and recent developments. *3 Biotech*. https://doi.org/10.1007/s13205-014-0246-5.

Sánchez, C. (2009). Lignocellulosic residues: Biodegradation and bioconversion by fungi. *Biotechnology Advances*. https://doi.org/10.1016/j.biotechadv.2008.11.001.

Sarkar, N., Ghosh, S. K., Bannerjee, S., & Aikat, K. (2012). Bioethanol production from agricultural wastes: An overview. *Renewable Energy*. https://doi.org/10.1016/j.renene.2011.06.045.

Sathitsuksanoh, N., Zhu, Z., & Zhang, Y. H. P. (2012). Cellulose solvent-based pretreatment for corn stover and avicel: Concentrated phosphoric acid versus ionic liquid [BMIM]Cl. *Cellulose*. https://doi.org/10.1007/s10570-012-9719-z.

Sebayang, A. H., Masjuki, H. H., Ong, H. C., Dharma, S., Silitonga, A. S., Mahlia, T. M. I., & Aditiya, H. B. (2016). A perspective on bioethanol production from biomass as alternative fuel for spark ignition engine. *RSC Advances*. https://doi.org/10.1039/c5ra24983j.

Sindhu, R., Binod, P., & Pandey, A. (2016). Biological pretreatment of lignocellulosic biomass – an overview. *Bioresource Technology*. https://doi.org/10.1016/j.biortech.2015.08.030.

Song, Y., Cho, E. J., Park, C. S., Oh, C. H., Park, B. J., & Bae, H. J. (2019). A strategy for sequential fermentation by *Saccharomyces cerevisiae* and *Pichia stipitis* in bioethanol production from hardwoods. *Renewable Energy*. https://doi.org/10.1016/j.renene.2019.03.032.

Song, L., Yu, H., Ma, F., & Zhang, X. (2013). Biological pretreatment under non-sterile conditions for enzymatic hydrolysis of corn stover. *BioResources*. https://doi.org/10.15376/biores.8.3.3802-3816.

Stanbury, P. F., Whitaker, A., & Hall, S. J. (2017). The recovery and purification of fermentation products. In *Principles of fermentation technology*. https://doi.org/10.1016/b978-0-08-099953-1.00010-7.

Taha, M., Shahsavari, E., Al-Hothaly, K., Mouradov, A., Smith, A. T., Ball, A. S., & Adetutu, E. M. (2015). Enhanced biological straw saccharification through coculturing of lignocellulose-degrading microorganisms. *Applied Biochemistry and Biotechnology.* https://doi.org/10.1007/s12010-015-1539-9.

Taherzadeh, M. J., & Karimi, K. (2008). Pretreatment of lignocellulosic wastes to improve ethanol and biogas production: A review. *International Journal of Molecular Sciences.* https://doi.org/10.3390/ijms9091621.

Travaini, R., Martín-Juárez, J., Lorenzo-Hernando, A., & Bolado-Rodríguez, S. (2016). Ozonolysis: An advantageous pretreatment for lignocellulosic biomass revisited. *Bioresource Technology.* https://doi.org/10.1016/j.biortech.2015.08.143.

Wagner, A. O., Janetschek, J., & Illmer, P. (2018). Using digestate compost as a substrate for anaerobic digestion. *Chemical Engineering and Technology.* https://doi.org/10.1002/ceat.201700386.

Xu, W., Fu, S., Yang, Z., Lu, J., & Guo, R. (2018). Improved methane production from corn straw by microaerobic pretreatment with a pure bacteria system. *Bioresource Technology.* https://doi.org/10.1016/j.biortech.2018.02.046.

Yusuf, A. A., Onu, P., Hassanb, A. S., Tunji, A.,L., Oyagbola, I. A., Mustafad, M. M., & Yusuf, D. A. (2019). Municipality solid waste management system for Mukono district. *Procedia Manufacturing, 35*, 613–622. https://doi.org/10.1016/j.promfg.2019.06.003.

Zabed, H., Sahu, J. N., Boyce, A. N., & Faruq, G. (2016). Fuel ethanol production from lignocellulosic biomass: An overview on feedstocks and technological approaches. *Renewable and Sustainable Energy Reviews.* https://doi.org/10.1016/j.rser.2016.08.038.

Zhao, X., Liu, W., Deng, Y., & Zhu, J. Y. (2017). Low-temperature microbial and direct conversion of lignocellulosic biomass to electricity: Advances and challenges. *Renewable and Sustainable Energy Reviews.* https://doi.org/10.1016/j.rser.2016.12.055.

Zheng, Y., Pan, Z., & Zhang, R. (2009). Overview of biomass pretreatment for cellulosic ethanol production. *International Journal of Agricultural and Biological Engineering.* https://doi.org/10.3965/j.issn.1934-6344.2009.03.051-068.

Zheng, Y., Zhao, J., Xu, F., & Li, Y. (2014). Pretreatment of lignocellulosic biomass for enhanced biogas production. *Progress in Energy and Combustion Science.* https://doi.org/10.1016/j.pecs.2014.01.001.

CHAPTER

Sustainable agricultural waste diversity: advances in green energy and materials production

4

4.1 Introduction

The notion that "one's trash is another man's treasure" is experienced in real-life scenarios, as new business networks are springing up to retrieve all manner of wastes, locally and internationally. Hence, the need for a sustainable approach to address the imminent menace facing resource depletion and environmental safety cannot be overemphasized, especially with regard to agricultural wastes (Onu & Mbohwa, 2018b, 2019c; Raji & Onu, 2017). A sustainable process or approach is a thermodynamic equilibrium situation that may only slightly change in the cause of the future due to a shift from what is considered as normal (Onu & Mbohwa, 2019a; Onu & Mbohwa, 2018a; Val del Río, Campos Gómez, & Mosquera Corral, 2016). A sustainable and unsustainable occurrence is a subjective concept due to the different conditions and variables that guide their actualization. For example, a sustainable strategy to manage waste generation through the landfills approach fails to remain sustainable whence the design can no longer handle further waste entries. The world Summit on Sustainable Development was charged in the early year 2000 to promote developing nations in areas of resource management and sustainable development approaches to control global depletion and distribution of resources, which infringe on the conservation of the global eco-system.

New motivations toward ways of waste elimination and reduction in generation and sustainable disposition that will promote climate change mitigation take its bearing from the Brundtland Report (Brundtland, 1987). However, there is not sufficient detail about the specifics, and models to actualize the goals, and, thus, has promulgated more research and investigations. Agricultural waste diversity is promising in diverse areas of application, one of which is to replace the overusage of nonbiodegradable synthetic or petroleum-based materials used as wrappers or for carriage (Onu & Mbohwa, 2019c). This chapter aims to demonstrate the routes and likeliness of agricultural waste to be used as bio-based products (automotive fuels) and materials, in the transport and food/agroindustrial sector as packaging or carriage material, respectively. Building on previous knowledge, this book shows

Agricultural Waste Diversity and Sustainability Issues. https://doi.org/10.1016/B978-0-323-85402-3.00008-5

the necessary steps and strategy that eventually leads to productive, with respect to bioproducts from agricultural wastes after the biomass pretreatment stages, and hence discusses sustainable initiatives in agricultural resources conversion and product designs.

4.2 Sustainable bioenergy production

Bioenergy is defined as "material which is directly or indirectly produced by photosynthesis and which is utilized as a feedstock in the manufacture of fuels and substitutes for petrochemical and other energy-intensive products" (IEA, 2019). The production and development of bioenergy and materials are based on the utilization and application of biorefinery (Fahd, Fiorentino, Mellino, & Ulgiati, 2012). Biorefinery has been defined by the International Energy Agency (IEA) (Technology roadmap: bioenergy for heat and power, 2012), as a facility that sustainably generates products of commercial interest using only biomass as substrate. The biorefinery concept can be thought of as the equivalent of an oil refinery where there are different materials derived from petroleum. In the case of a biorefinery, the feedstock is biomass, which includes the agricultural, food industry, and municipal organic waste residues, among others (FitzPatrick, Champagne, Cunningham, & Whitney, 2010). Hence, by combining different strategies and technologies, they are transformed mainly into energy: heat, electricity or biofuels, and other biomaterials with high added value in the market (Lawal et al., 2019; Yusuf et al., 2019; Raji & Onu, 2017; Urbaniec & Bakker, 2015). The use of biomass as an alternative clean energy source compared with nonrenewable resources involves a series of sequential processes, including preparation and processing into products of interest (Fig. 4.1A and B), with inference to how much waste is produced. However, much emphasis has been given to the issue of logistical consideration and infrastructural requirement as well as integration and innovative interactions to lead raw materials handling and biomass conversion for gainful bioenergy production. Regardless, the chances of land scarcity and competition for space for agrowaste management or enforcement of the policies to support public practice will require a great deal of education and sensitization on the opportunities of sustainable bioenergy production.

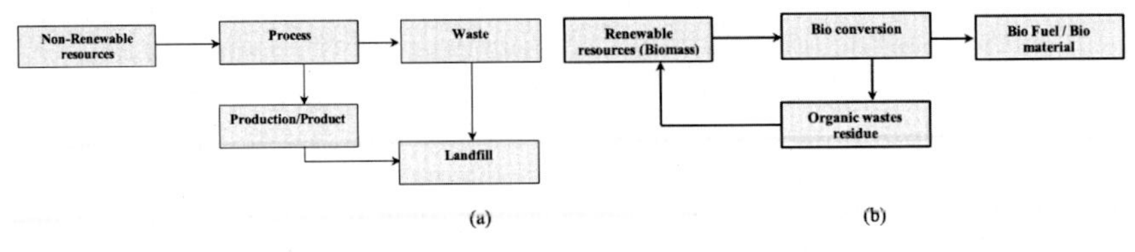

FIGURE 4.1

Potential for sustainable production from renewable resources versus nonrenewable resources.

Biofuel production and areas of application

Biomass conversion to fuels is summarized into three main techniques, with the technological processes illustrated in Fig. 4.2. Thus, it encompasses the

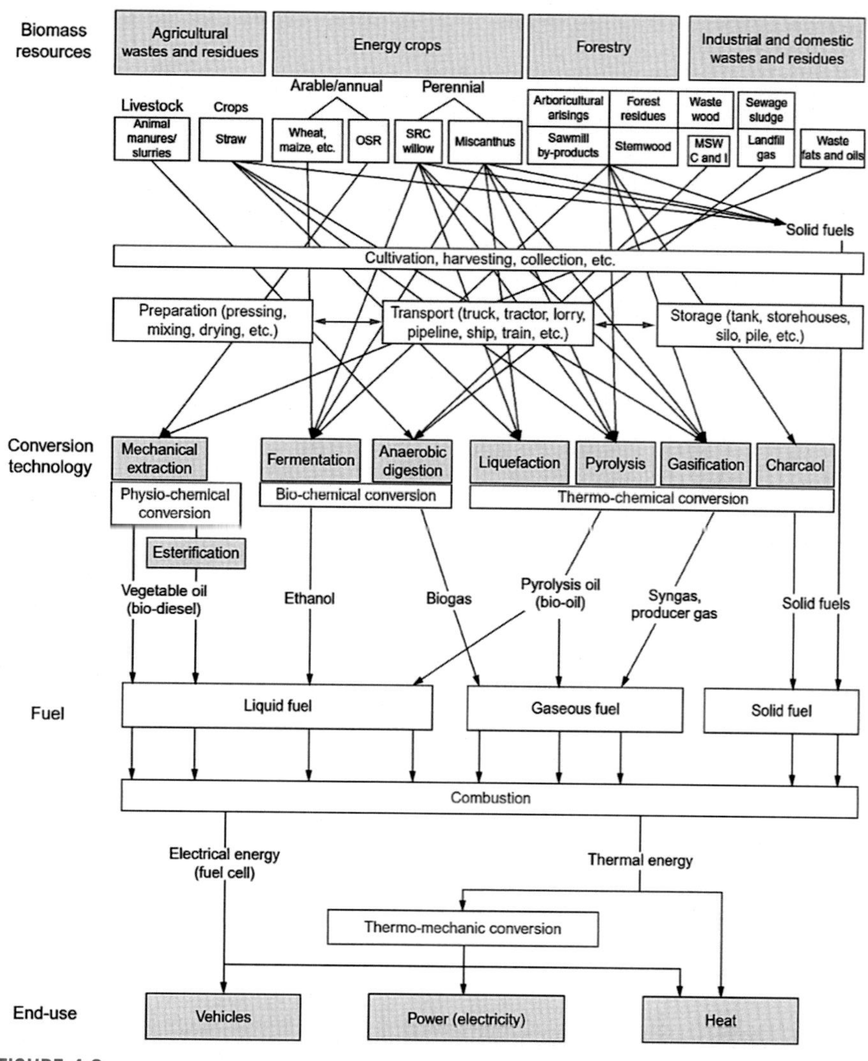

FIGURE 4.2

Schematic representation of biomass conversion pathways.

Adapted for Adams, P., Bridgwater, T., Lea-Langton, A., Ross, A., & Watson, I. (2017). Biomass conversion technologies. Report to NNFCC. In: Greenhouse gas balances of bioenergy systems. *https://doi.org/10.1016/B978-0-08-101036-5.00008-2. Biomass conversion technologies.* Green House Balance of Bioenergy Systems, 2018*:107–137.*

physicochemical, biochemical, and thermo-chemical conversion processes that follow the production of mechanically derived vegetable oil, ethanol, biogas, biooil, syngas, and other forms of fuels (solid). In the combustion conversion process, the energy release is direct and sufficient to power engines via a boiler and turbine arrangement. The performances of the aforementioned process can be improved through optimization of the operation process, assumed to have the same efficiency as fossil fuel-based or the natural gas resources (Onu & Mbohwa, 2018b). The physicochemical process and production of vegetable oils and also the transesterification process yield biodiesel production which is vital and used as blends with traditional mineral oils to improve cetane number, leading to excellent lubricity (Pourzolfaghar, Abnisa, Daud, & Aroua, 2016; Sutanto et al., 2017; Taherzadeh & Karimi, 2008). The by-product of the alcoholic fermentation process (bioethanol) of biomasses is essential as an octane enhancer for petroleum blends (Iakovou, Karagiannidis, Vlachos, Toka, & Malamakis, 2010; Mohd & Aziz, 2016), whereas biogas after collection from the anaerobic digestion process leaves digestate that is useful for promoting soil characteristics or dried out to form lightweight composite materials.

Anaerobic digestion is a process that takes place in a reactor and leads to the formation of methane gas (bio) that can be used for different purposes, like in heating, cooking, and in more recent cases to produce welding flames. The nature of wastes targeted for the anaerobic digestion process takes the wet form of wastes, such as biodegradable discards from homes and large farms (McCrone, Moslener, D'Estais, & Grünig, 2017). This process effectively supports the act of composting, especially of food and agricultural products, thus resolving the discomfort from the foully smells in open-air disposal (Adekunle & Okolie, 2015; Mao, Feng, Wang, & Ren, 2015).

The thermochemical liquefaction process is one that promotes the yield of biooil; however, the cost of the technology poses a challenge to technique. The pyrolysis process yields biooil, syngas, and charcoal from biomass conversion with the help of a catalyst to facilitate the operation and promote increased production. Although syngas (from gasification process) is considered a low calorific gas, it is mostly used as an upgrade or directly to power hydrogen fuel cells (Arun, Sharma, & Dalai, 2015; Onu, Abolarin, & Anafi, 2017; Shedrack et al., 2019). The torrefaction process (conversion of biomass to charcoal) produces biocoal. This occurs during the burning of biomass below a temperature of 320°C, in the absence of oxygen, and having significant decomposition effect on cellulose, hemicellulose, and lignin constituent of the biomass (Nunes, Matias, & Catalão, 2014).

Several conditions exist that can serve as a hindrance to waste conversion projects. The different processes includes organizing and preparing the desired biomass wastes; transport protocol; means to generate the energy required, and the specific operational techniques to be applied, which in most case are usually costly, painstaking, and time-consuming (Aziz, Hanafiah, & Gheewala, 2019; Bajwa, Peterson, Sharma, Shojaeiarani, & Bajwa, 2018; Paudel et al., 2017). Moreover, the development of bio-based and biodegradable products from wastes offers a new dimension in agrowaste management. Manufacturing products such as kitchen utensils, containers, and bottles, to mention a few, narrows the application of biomaterials (biodegradable polymers) for green materials or disposables products (Johansson et al., 2012). Also, while natural biomaterials have found several usages in the medical sector, the need to extend further research into sustainable means of developing everyday products that are environmentally unsafe is paramount.

Biomass-derived from plants, i.e., feedstocks for the production of clean fuels, has also been categorized into three: starch, including sugar, triglyceride, and lignocellulosic matter. Starch is a white powder (physical appearance) consisting of polysaccharide in which glucose monomers linked in alpha (α)-1,4 chains (Payne, 2009). The starch is produced commercially from corn, beets, and sugarcane. Chemical structure of triglycerides consists of linkage of glycerol to three fatty acid molecules, and thus, these are known as triesters. Triglycerides are present in vegetables and animals as body fat (Ferretti, Spotti, & Di Cosimo, 2018). While lignocellulosic feedstocks are leftovers or dry matter of crops, which remained after processing farm produce or algal biomass (Zhang et al., 2010), they contain hemicellulose by a proportion of 25%–30%, the cellulose of 40%–50%, and lignin of 15%–20%. The hemicellulose and cellulose are carbohydrate polymers that are made of different sugar monomers of six and five carbons and firmly bonded to lignin (Amarasekara, 2013).

Lignocellulosic biomass resources are cheap and extensively available; they are residues that have no effect on the food supply chain. Also, they are efficient for the production of biodiesels through the esterification process. These materials have a great potential to replace carbon resource as they can be processed to produce fuels for automotive and other valuable chemicals (Dai, Liu, & Si, 2018; Delidovich, Leonhard, & Palkovits, 2014; Mariscal, Maireles-Torres, Ojeda, Sádaba, & López Granados, 2016).

Lignocellulose conversion process in brief

The production of ethanol from starch is carried out through the hydrolysis and fermentation route. Triglyceride derived from vegetable oils such as soya and sunflower oils and fatty acid contents can be converted to biodiesel by esterification process. Biodiesel is mixable with fossil fuel (diesel) for direct use in diesel engines, owing to the cellulose content in lignocellulosic biomass that can be converted into hexamethyl furan (HMF) and levulinic acid (LA). Also, hemicellulose derived from lignocellulose material can be processed to produce furfural, while the lignin content facilitates the production of biooils. All these chemicals can be converted into liquid alkanes, which can be utilized as transportation fuels (Bozell & Petersen, 2010).

Lignocellulosic biomass conversion to biofuel may be conducted in two different approaches. The first, which involves the thermochemical process, comprises three activities: gasification, pyrolysis, and liquefaction. The gasification process leads to the production of syngas, which can be converted into compatible diesel or gasoline fuels. Pyrolysis and liquefaction lead to biooil production. These biooils can be upgraded to compatible transport fuels. The second approach is a hydrolysis process, where at least two different catalytic conversions are possible, one of which results in sorbitol formation, and another results in hexamethyl furan (Furfural). Sorbitol results in monofunctional groups, which can, in turn, be convertible into gasoline/diesel via ketonization or aldol condensation. HMF can also be converted to levinic acid and sequentially to gamma-valerolactone (GVL or γ-valerolactone), which is a potential fuel and green solvent.

Furfural is a first-stage chemical produced from lignocellulosic biomass. This chemical can be utilized for platform chemicals such as furfuryl alcohol, 2-methylfuran, tetrahydrofuran, cyclopentanone, and γ-valerolactone (Bui, Luo, Gunther, & Román-Leshkov, 2013; Yan, Wu, Lafleur, & Jarvis, 2014). Its production is facilitated by either the homogeneous and heterogeneous catalysts. During the homogeneous process, different biomass undergoes a chemical conversion, at 180°C for the duration of 30 minutes, during hydrolysis of corncob in the presence of a catalyst: acetic acid and $FeCl_3$ resulted in the furfural production, with up to 67.89% conversion. The separation rates of lignin and cellulose have been observed to reach 54.79% and 25.71%, respectively, according to Mao et al. (Mao, Zhang, Gao, & Li, 2012). Also, tests for the effectiveness of niobium phosphate (NbP) catalyst for xylose dehydration at 160°C within a process duration of 30 minutes had yielded 43% furfural from corn stover. Furthermore, the catalyst was reemployed two times and observed the good conversion of furfural nearer to the first yield observed (Gómez Bernal, Bernazzani, & Raspolli Galletti, 2014). In the heterogeneous process, the carbon solid acid catalyst is employed in the production of furfural from xylose and corn stalk, in the presence of a sulfonating substance (4-benzene diazonium sulfonate) at 200°C within a process duration of 100 minutes, which had yielded 78.5% and 60.6% of furfural from xylose and corn stalk, respectively, with inference to the characteristic of the catalyst to be recyclable (Zhang et al., 2016).

Hydrogenation of furfural leads to the furfuryl alcohol, which is a useful chemical in the production of biofuels and polymers such as rubbers, resins, and fibers (Li, Jia, & Wang, 2016; Mariscal et al., 2016). The hydrogenation process, while in the vapor or liquid phase, is facilitated by transition metals (Gong et al., 2017). Also, the use of Pd–Cu (5% and 5%) catalyst carried by MgO and $Mg(OH)_2$, and water as a solvent, at 110°C with a hydrogen gas pressure of 0.6 MPa, has been reported with 100% conversion of furfural and 98% of alcohol (Fulajtárova et al., 2015). Also, the investigation into the use of platinum nanoparticle (γ-Al_2O_3, CeO_2, and MgO) of 4 nm, as a catalyst for hydrogenation of furfural (liquid phase), gave excellent outcomes, with the catalysts being recyclable and having to sustain conversion efficiency, and thus increased as the particle size further decreased (Taylor et al., 2016). The use of Cu as a catalyst during the hydrogenation of furfural using amphoteric metal oxide increases the catalytic activity and enhances the activation centers on the Cu surface by modifying the electron density causing desorption of furan and slowing deactivation. Hence, in a conversion process, lasting up to 5 hours, and not below 210°C, 80% conversion of furan is tenable and is composed of two products:

furfuryl alcohol, which is an insignificant portion, and 2-methyl furan in minute quantity. The assigned reason for the formation of 2-methyl furan is due to the presence of acid sites that caused the hydrogenolysis of furfuryl alcohol (Guerrero-Torres et al., 2019).

4.3 Sustainable biomaterials: selection, design, and production

The usage of plastics and thin-film metal of synthetic materials continue to pose a great risk and serious health and environmental risks. Regardless, food carriage producers continue to utilize petrochemicals to design products. The application of these products implies continuous exploration of the fossils (oil), thus, engaging in high costs, energy-intensive, and environmentally unsafe activities that endanger the sustenance of the world's ecosystem. Contrarily, the use of natural biodegradable materials for packaging and carriage purposes attracts research interest and the sensitization of active players to explore its opportunities (Vroman & Tighzert, 2009). These proposals are supported through the redesign of packaging materials to become recyclable through the adoption of unwanted agricultural resources in the production of natural biodegradable products. The promulgation and utilization of such materials will lead to a reduction in the exploitation of fossil resources and contribute to the decrease of CO_2 release to the atmosphere (Molenveld, van den Oever, & Bos, 2015).

An investigation into the development of polymer from agrowastes, occurring from farm discards, municipal and households' wastes, and industrial wastes is timely to promote green products creation. The rising interest around the world on the topic of organic materials and the evolution of new processes and pathways for it to be commercially viable will, in turn, reduce the cost of the process and lead to widespread acceptance of the concept (Castro-Aguirre, Iñiguez-Franco, Samsudin, Fang, & Auras, 2016; Onu & Mbohwa, 2019b; Onu & Mbohwa, 2018a). The goal is to have packaging materials and carriers that are sustainable, cost-effective, quality, and ecofriendly (Johansson et al., 2012), having lightweight, with excellent mechanical and thermal properties (high strength and thermal adjustment) (Lü, Ye, & Liu, 2009).

Recent research on bioproducts creation focuses on the production of biomaterials such as enzymes, biopolymers, biosolvents, biosurfactants, organic acids, and pigments. In this chapter, investigations surrounding the production of these biomaterials by fermentation over the past 5 years are discussed, highlighting the product yield from each process. Currently, there are several biotechnological alternatives, developed as part of a sustainable system that allows the recovery of agrowastes, which can subsequently be transformed into a variety of products with commercial value (Galanakis, 2012). Agricultural waste is conveniently exploited for biomaterial production due to the presence of pigments, polyphenolic compounds, essential oils, and fibers, which can be used in diverse purposes. Biomaterial production from agrowaste can be achieved through the extraction or fermentation processes, either with pretreatment to obtain fermentable sugars or without pretreatment by solid-state fermentation (Fig. 4.3).

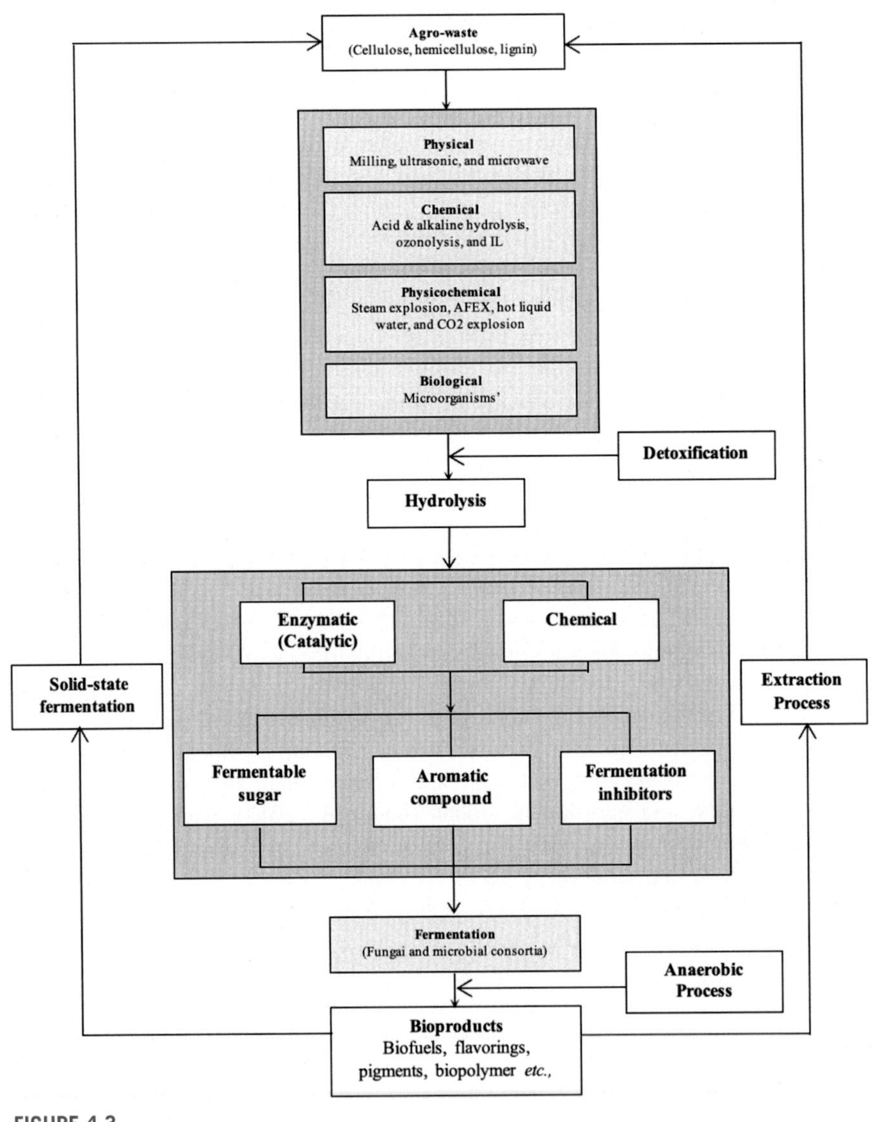

FIGURE 4.3

Simplified illustration of the valorization of agricultural waste to obtain biofuels and biomaterials.

The rich polysaccharide nature of starch, derived from the varieties of agrowaste, offers it the potential to be used as a polymer for the development of nascent materials that can withstand the challenge of synthetic nonbiodegradable materials mostly used for packing or carrying purposes. More so, depending on the area of application, as per weight, or lifting of nonregularized shapes and dimensions, several pretreatment or chemical changes (plasticization) are considered (Ferreira, Alves, & Coelhoso, 2016). Starch (edible) and its blends are being considered for food packaging applications (Tabasum et al., 2018). Also, observation shows that the contribution in energy consumption for the production of starch-based plastic pellets against polyethylene is lesser (Marina Patricia Arrieta, Samper, Aldas, & López, 2017).

Although several wastes have been used as substrates, in recent times, the focus is mainly on the use of waste from commodities, which have high rates of production worldwide, such as sugar cane, maize, rice, and wheat (biomasses). These biomasses are an important source of cellulose, hemicellulose, and lignocellulose and mostly used feedstock (with significantly, high starch content) for producing fuels, chemicals, and other industrially manufactured products. These fall in the same categories as nonbiodegradable bioplastics (of agricultural source renewables) or by-products of the fermentation processes, leading to the production of polymers. The formation of lactic acid (single monomer), which is a base of the polylactide (PLA), features favorable mechanical properties, lightweight, and transparent characteristics, preferable to thermoplastics such as polystyrene and polyethylene terephthalate. PLA is viable through the transformation operations and conversion to varieties of products, having less carbonaceous effect from emission and greenhouse gas releases compared with petroleum-based polymers, according to Jamshidian, Tehrany, Imran, Jacquot, and Desobry (2010).

It is worth to note that the application of PLA is fast developing and has been adopted for food wraps and ready-to-consume meal packs. Another biodegradable thermoplastic polymer of the aliphatic polyesters base is the polyhydroxyalkanoates (PHAs), which unlike the polylactide feature great thermomechanical abilities, also produced from varieties of microorganism after the fermentation process. The PHA is truculent and mostly utilized for the production of handy low-weight carriages. They are soluble in PHA and promote excellent physical, thermal, and mechanical properties (Arrieta, Samper, López, & Jiménez, 2014).

Biopolymer production process in brief

Studies have presented the possibility of using agrowastes as substrates for the production of biopolymers (Table 4.1). Exopolysaccharides (EPS) produced by fungi and bacteria (such as lactic acid bacteria, LAB) have antitumor, hypoglycemic, and immunostimulating activity. It was shown that EPS produced by *Inonotus obliquus* from ground corn stover had antioxidant activity using 2,2-diphenyl-1-picrylhydrazyl radical (DPPH) with an IC50 (mg/mL) between 1.32 and 4.35 (Xiang, Xu, & Li, 2012). Similarly, EPS synthesized by *Bacillus megaterium* RB-

Table 4.1 Biopolymers produced by microorganisms using agroindustrial wastes as a carbon source.

Residue	Producing microorganism	Biopolymer	Yield	Reference
Corn cob	Inonotus obliquus (CBS314.39)	Exopolysaccharide	N/A	Xiang et al. (2012)[a,b]
Jute	*Bacillus megaterium* RB-05	Exopolysaccharide	0.297 g/g	Roy Chowdhury et al. (2011)
Rice straw	*Bacillus cereus* PS 10	Poly-3-hydroxybutyrate	10.61 g/L	Cesário et al. (2014) [b]
Rice straw	*Bacillus firmus* NII 0830	Poly-3-hydroxybutyrate	0.16 g/L	Sindhu et al. (2013)
Rice straw	*Bacillus sphaericus* 0838	Poly-3-hydroxybutyrate	0.08 g/L	Sindhu et al. (2013)
Chicken feathers	*Morchella esculenta*	Exopolysaccharide	4.8 g/L	Taskin et al. (2012)
Used palm oil	*Cupriavidus necator*	Poly-3-hydroxybutyrate	0.8 g/g	Rao et al. (2010)
Frying oil	*C. necator* H16 NCIMB10442	Poly-3-hydroxybutyrate	1.2 g/L	Verlinden et al. (2011)
Wheat cob	*Burkholderia sacchari* DSM 17165	Poly-3-hydroxybutyrate	0.22 g/g	Cesário et al. (2014)
Tequila bagasse	*Saccharophagus degradans* ATCC 43961	Polyhydroxyalkanoates	1.5 g/L	Munoz & Riley (2008)
Jatropha	*Pseudomonas oleovorans* ATCC 29347	Polyhydroxyalkanoates	26.06 g/g	Allen, Anderson, Ayorinde, & Eribo (2010)
Coconut water	*Lactobacillus confusus*	Exopolysaccharides	18 g/L	Seesuriyachan et al.(2010) [a]
Cane juice	*L. confusus*	Exopolysaccharides	62 g/L	Seesuriyachan et al. (2010) [a]
Frying oil	*C. necator* H16 NCIMB10442	Poly-3-hydroxybutyrate	1.2 g/L	Xiang et al. (2012) [a,b]

[a] *Solid-state fermentation.*
[b] *Statistical optimization for production.*

05 from jute wastes as carbon source were evaluated as potential antioxidants (Roy Chowdhury, Kumar Basak, Sen, & Adhikari, 2011). Fungi, such as *Morchella esculenta*, are also effective EPS producers, and this microorganism was able to produce 54% more biopolymers adding 10 g/L of chicken feather peptone compared with basic medium (only yeast and glucose) (Taskin, Ozkan, Atici, & Aydogan, 2012).

There are six main monomers used as building blocks for biopolymer synthesis: hydroxyalkanoates, lactic acid, succinic acid, ethylene, 1,3-propanediol, and cis-3,5-

cyclohexadien-1,2-diols, which are synthesized in vitro, and PHA, which is synthesized in vivo. Other monomers are polymerized in vitro to form poly(lactic acid) (PLA), poly(butylene succinate) (PBS), polyethylene (PE), poly(trimethylene terephthalate) (PTT), and poly(p-phenylene) (PPP) (Chen, 2010). PHAs are synthesized for carbon and energy storage in response to stress or excess of carbon sources and the lack of at least one essential nutrient such as nitrogen, oxygen, or phosphorus. The functional properties of the PHA depend, among other factors, on the type of fatty acid, specifically aliphatic chain length and the presence of unsaturation. PHAs are more commonly composed of 3-hydroxy alkanoates with a variation on length and composition, which produce a wide range of potential applications (Tortajada, da Silva, & Prieto, 2013). Poly-B-hydroxybutyrate (PHB), the most studied PHA, is similar to polypropylene regarding its low oxygen permeability and thermoplastic properties, which largely determine its application potential, but because of its high production cost, the commercialization on a large scale has not been possible (Morais et al., 2014).

Like other biomaterials, it is possible to reduce their costs of production using the waste from agricultural farming activities as substrates and to optimize the factors affecting the particular biopolymer production. Species from *Bacillus* genus are among the most frequently used microorganisms for PHB production (Table 4.1). Sharma and Bajaj (2015) determined that the pH of the medium has a greater positive effect on PHB production by *Bacillus cereus* PS 10 from the fermentation of rice waste, although the influence of NH4Cl and the substrate concentration is also observed. Differences in the capacity to produce PHB have been described for two *Bacillus* species (Table 4.1). While *Bacillus firmus* NII 0830 showed greater capacity to produce PHB than *Bacillus sphaericus* 0838 from rice straw, the longer incubation time was necessary to achieve a better performance since the synthesis of PHB occurs in the stationary phase of bacterial growth (Sindhu, Silviya, Binod, & Pandey, 2013).

Other bacteria's that exhibit similar characteristics include, but are not limited to *Paracoccus denitrificans*, *Cupriavidus necator*, *Comamonas testosteroni*, and *Pseudomonas* sp. (Sawant, Salunke, & Kim, 2015). Cavalheiro et al. (2012) determined that an increase in dissolved oxygen of *C. necator* culture medium, using glycerol waste as carbon source, doubled the yield of PHB. The addition of fat or oil residues favors the production of PHB, and generally of PHA, as fatty acids obtained by enzymatic hydrolysis are metabolized via β-oxidation for PHA synthesis. Using 20% of palm oil, the yield of PHB synthesis by *C. necator* (Table 4.1) was 0.8 g/g at 144 h (Rao, Sridhar, & Sehgal, 2010), but when using waste frying oil at the same concentration, the yield of PHB was increased to 1.2 g/L after 72 hours. In contrast, when using pure vegetable oil, 50% less PHB was obtained (Verlinden et al., 2011). The production of these polymers, regardless of their origin, is also dependent on culture medium conditions. Seesuriyachan, Techapun, Shinkawa, and Sasaki (2010) evaluated the production of EPS by *Lactobacillus confusus*, by fermentation of coconut water and cane juice, obtaining yields of 62 g/L for cane juice and 18 g/L for coconut water. But, when the nitrogen source was reduced fivefold, the EPS synthesis increased to 62 g/L when using the cane juice as a substrate.

Bioplastics used as films, coating, and materials for packaging purposes

The application of biomaterials from wastes resource utilization meets the production of common everyday usage, ranging from packing to wrappings or coating (edible films), and can facilitate food preservation (Okcu, Yavuz, & Kerse, 2018). Also, the utilization of the polysaccharides and proteins-based polymers and their physical capabilities has been explored for edibility potential (Espitia, Du, Avena-Bustillos, Soares, & McHugh, 2014; Jiménez, Fabra, Talens, & Chiralt, 2012; Ocak, 2019). The use of polysaccharides for food coating is faced with the challenge of moisture sensitivity and poor water retaining capability. However, this is a preferable property for gaseous exchange tendencies. Moreover, blending these polymers improves their hydrophobic characteristics. Nonetheless, the protein-based biopolymers have been researched to have proven qualities also that surpass the polysaccharides with sustainable properties for use as an edible film and food wraps or packaging materials, having high physical, chemical, and enzymatic performance (Cerqueira et al., 2011; Han, 2013; Sharif et al., 2018), utilized for milk and dairy containers, juice, and water bottles and for packaging vegetable produce. Optimization of the biopolymers is recommended by the addition of plasticizers, which is optimally controlled, and the selection criteria being dependent on their ability to permit water permeability and gaseous exchange (Ocak, 2019).

The application of biomaterials of agrowaste sources in the food processing and manufacturing industries has continued for a while without the awareness of the consumers who patronize some of the products. The application has been used to enhance the appearance of fruits and vegetables, as it helps to improve their life span, viz., the coating (edible coats) (McClements, Bai, & Chung, 2017). Moreover, in the pharmaceutical companies, nowadays these coats are used to envelop and preserve essentials that are meant for, and dispensed to patients, as per medicines (Trinetta & Cutter, 2016). Wastes from different vegetables and fruits have been developed into edible films to provide unique flavorings for vegetarians, or as wraps, for convenience of eating different types of seafood (Liu, Ma, Gao, & McClements, 2017). Edible cups and containers (single use) are fascinatingly the muse of the future and will become widespread, replacing the synthetic materials being used currently, thus combating the ill effects of excessive exploitation of crude oils or the by-products thereof that threaten livelihood. Biodegrading packages have been experimented, with the use of edible caramelized sugar coating, the agar—agar seaweed, and beeswax, for serving oil products, smoothies, and rice to mention a few (Tomorrowmachine.se, 2013).

4.4 Conclusions and prospects

Conclusively, nascent techniques and technologies in the production of clean fuels and biomaterial from agrowaste have been presented. This contributes to the status of developing countries to achieve the 17 Sustainable Development Goals, which, as such, address resource depletion and environmentally unsafe practices in Africa. The authors proffer that the few cases and processes inferred in this study can be optimized to increase product yields through the genetic modification of microorganisms or using statistical design. The researched techniques (for biofuel and ester production) are proven to be sustainable strategies in terms of reducing waste from an agricultural operation and the creation of eco-friendly products. The promotion of feasible regulatory policies to guide the design and production of the material from agricultural resources conversion is paramount to propel innovation in biofuel and bioplastic production. However, biopolymer materials are emerging as packaging material to protect against environmental commitment and reduce waste from agricultural activities.

The availability of multiple and efficient techniques and processes for bioenergy production or the routes for biomaterial development will require the intervention and interests of stakeholders to facilitate and promote the processes. Moreover, bioenergy production from the biomass resources or agricultural waste products must overcome the limitations that prevent its acceptability as a potential for a country's energy mix and integration of the technology to the national grid. Future intervention is expected for improving heat saleability and the printability of biodegradable films. Additionally, how to make green products cost-competitive with conventional plastics? Furthermore, the exploration into microstructural and physicochemical properties and biodegradability of bio-based packaging materials, as well as efforts to propagate their areas of application, is recommended. Increasing consumer awareness and acceptance of such innovation in food packaging will springboard the concept and lead to the commercialization of the idea.

References

Adams, P., Bridgwater, T., Lea-Langton, A., Ross, A., & Watson, I. (2017). Biomass conversion technologies. Report to NNFCC. In *Greenhouse gas balances of bioenergy systems*. https://doi.org/10.1016/B978-0-08-101036-5.00008-2.

Adekunle, K. F., & Okolie, J. A. (2015). A review of biochemical process of anaerobic digestion. *Advances in Bioscience and Biotechnology*. https://doi.org/10.4236/abb.2015.63020.

Allen, A. D., Anderson, W. A., Ayorinde, F. O., & Eribo, B. E. (2010). Biosynthesis and characterization of copolymer poly(3HB-co-3HV) from saponified *Jatropha curcas* oil by *Pseudomonas oleovorans*. *Journal of Industrial Microbiology and Biotechnology*. https://doi.org/10.1007/s10295-010-0732-7.

Amarasekara, A. S. (2013). Acid hydrolysis of cellulose and hemicellulose. In *Handbook of cellulosic ethanol*. https://doi.org/10.1002/9781118878750.ch7.

Arrieta, Marina Patricia, Samper, M. D., Aldas, M., & López, J. (2017). On the use of PLA-PHB blends for sustainable food packaging applications. *Materials*. https://doi.org/10.3390/ma10091008.

Arrieta, Marina P., Samper, M. D., López, J., & Jiménez, A. (2014). Combined effect of poly(-hydroxybutyrate) and plasticizers on polylactic acid properties for film intended for food packaging. *Journal of Polymers and the Environment*. https://doi.org/10.1007/s10924-014-0654-y.

Arun, N., Sharma, R. V., & Dalai, A. K. (2015). Green diesel synthesis by hydrodeoxygenation of bio-based feedstocks: Strategies for catalyst design and development. *Renewable and Sustainable Energy Reviews*. https://doi.org/10.1016/j.rser.2015.03.074.

Aziz, N. I. H. A., Hanafiah, M. M., & Gheewala, S. H. (2019). A review on life cycle assessment of biogas production: Challenges and future perspectives in Malaysia. *Biomass and Bioenergy*. https://doi.org/10.1016/j.biombioe.2019.01.047.

Bajwa, D. S., Peterson, T., Sharma, N., Shojaeiarani, J., & Bajwa, S. G. (2018). A review of densified solid biomass for energy production. *Renewable and Sustainable Energy Reviews*. https://doi.org/10.1016/j.rser.2018.07.040.

Bozell, J. J., & Petersen, G. R. (2010). Technology development for the production of biobased products from biorefinery carbohydrates — the US Department of Energy's "top 10" revisited. *Green Chemistry*. https://doi.org/10.1039/b922014c.

Brundtland, G. H. (1987). *Our common future (Brundtland report)*. United Nations Commission. https://doi.org/10.1080/07488008808408783.

Bui, L., Luo, H., Gunther, W. R., & Román-Leshkov, Y. (2013). Domino reaction catalyzed by zeolites with Brønsted and Lewis acid sites for the production of γ-valerolactone from furfural. *Angewandte Chemie International Edition*. https://doi.org/10.1002/anie.201302575.

Castro-Aguirre, E., Iñiguez-Franco, F., Samsudin, H., Fang, X., & Auras, R. (2016). Poly(lactic acid)—mass production, processing, industrial applications, and end of life. *Advanced Drug Delivery Reviews*. https://doi.org/10.1016/j.addr.2016.03.010.

Cavalheiro, J. M. B. T., Raposo, R. S., de Almeida, M. C. M. D., Teresa Cesário, M., Sevrin, C., Grandfils, C., & da Fonseca, M. M. R. (2012). Effect of cultivation parameters on the production of poly(3-hydroxybutyrate-co-4-hydroxybutyrate) and poly(3-hydroxybutyrate-4-hydroxybutyrate-3-hydroxyvalerate) by cupriavidus necator using waste glycerol. *Bioresource Technology*. https://doi.org/10.1016/j.biortech.2012.01.176.

Cerqueira, M. A., Bourbon, A. I., Pinheiro, A. C., Martins, J. T., Souza, B. W. S., Teixeira, J. A., & Vicente, A. A. (2011). Galactomannans use in the development of edible films/coatings for food applications. *Trends in Food Science and Technology*. https://doi.org/10.1016/j.tifs.2011.07.002.

Cesário, M. T., Raposo, R. S., Almeida, M. C. M. D. D., van Keulen, F., Ferreira, B. S., Telo, J. P., & Fonseca, M. M. R. D. (2014). Production of poly(3-hydroxybutyrate-co-4-hydroxybutyrate) by *Burkholderia sacchari* using wheat straw hydrolysates and gamma-butyrolactone. *International Journal of Biological Macromolecules*. https://doi.org/10.1016/j.ijbiomac.2014.04.054.

Chen, G. Q. (2010). Introduction of bacterial plastics PHA, PLA, PBS, PE, PTT, and PPP, in plastics from bacteria, natural functions and applications. In *Microbiology monographs*. https://doi.org/10.1007/978-3-642-03287_5_3.

Dai, L., Liu, R., & Si, C. (2018). A novel functional lignin-based filler for pyrolysis and feedstock recycling of poly(l-lactide). *Green Chemistry*. https://doi.org/10.1039/c7gc03863a.

Delidovich, I., Leonhard, K., & Palkovits, R. (2014). Cellulose and hemicellulose valorisation: An integrated challenge of catalysis and reaction engineering. *Energy and Environmental Science*. https://doi.org/10.1039/c4ee01067a.

Espitia, P. J. P., Du, W. X., Avena-Bustillos, R. de J., Soares, N. de F. F., & McHugh, T. H. (2014). Edible films from pectin: Physical-mechanical and antimicrobial properties — a review. *Food Hydrocolloids*. https://doi.org/10.1016/j.foodhyd.2013.06.005.

Fahd, S., Fiorentino, G., Mellino, S., & Ulgiati, S. (2012). Cropping bioenergy and biomaterials in marginal land: The added value of the biorefinery concept. *Energy*. https://doi.org/10.1016/j.energy.2011.08.023.

Ferreira, A. R. V., Alves, V. D., & Coelhoso, I. M. (2016). Polysaccharide-based membranes in food packaging applications. *Membranes*. https://doi.org/10.3390/membranes6020022.

Ferretti, C. A., Spotti, M. L., & Di Cosimo, J. I. (2018). Diglyceride-rich oils from glycerolysis of edible vegetable oils. *Catalysis Today*. https://doi.org/10.1016/j.cattod.2017.04.008.

FitzPatrick, M., Champagne, P., Cunningham, M. F., & Whitney, R. A. (2010). A biorefinery processing perspective: Treatment of lignocellulosic materials for the production of value-added products. *Bioresource Technology*. https://doi.org/10.1016/j.biortech.2010.06.125.

Fulajtárova, K., Soták, T., Hronec, M., Vávra, I., Dobročka, E., & Omastová, M. (2015). Aqueous phase hydrogenation of furfural to furfuryl alcohol over Pd-Cu catalysts. *Applied Catalysis A: General*. https://doi.org/10.1016/j.apcata.2015.05.031.

Galanakis, C. M. (2012). Recovery of high added-value components from food wastes: Conventional, emerging technologies and commercialized applications. *Trends in Food Science and Technology*. https://doi.org/10.1016/j.tifs.2012.03.003.

Gómez Bernal, H., Bernazzani, L., & Raspolli Galletti, A. M. (2014). Furfural from corn stover hemicelluloses. A mineral acid-free approach. *Green Chemistry*. https://doi.org/10.1039/c4gc00450g.

Gong, W., Chen, C., Zhang, Y., Zhou, H., Wang, H., Zhang, H., … Zhao, H. (2017). Efficient synthesis of furfuryl alcohol from H_2-hydrogenation/transfer hydrogenation of furfural using sulfonate group modified Cu catalyst. *ACS Sustainable Chemistry and Engineering*. https://doi.org/10.1021/acssuschemeng.6b02343.

Guerrero-Torres, A., Jiménez-Gómez, C. P., Cecilia, J. A., García-Sancho, C., Quirante-Sánchez, J. J., Mérida-Robles, J. M., & Maireles-Torres, P. (2019). Influence of the incorporation of basic or amphoteric oxides on the performance of Cu-based catalysts supported on sepiolite in furfural hydrogenation. *Catalysts*. https://doi.org/10.3390/catal9040315.

Han, J. H. (2013). Edible films and coatings: A review. In *Innovations in food packaging* (2nd ed.). https://doi.org/10.1016/B978-0-12-394601-0.00009-6.

Iakovou, E., Karagiannidis, A., Vlachos, D., Toka, A., & Malamakis, A. (2010). Waste biomass to-energy supply chain management: A critical synthesis. *Waste Management*. https://doi.org/10.1016/j.wasman.2010.02.030.

IEA. (2019). *Sustainability of bioenergy | bioenergy*. Retrieved September 7, 2019, from https://www.ieabioenergy.com/about/.

Jamshidian, M., Tehrany, E. A., Imran, M., Jacquot, M., & Desobry, S. (2010). Poly-Lactic Acid: Production, applications, nanocomposites, and release studies. *Comprehensive Reviews in Food Science and Food Safety*. https://doi.org/10.1111/j.1541-4337.2010.00126.x.

Jiménez, A., Fabra, M. J., Talens, P., & Chiralt, A. (2012). Edible and biodegradable starch films: A review. *Food and Bioprocess Technology*. https://doi.org/10.1007/s11947-012-0835-4.

Johansson, C., Bras, J., Mondragon, I., Nechita, P., Plackett, D., Šimon, P., … Aucejo, S. (2012). Renewable fibers and bio-based materials for packaging applications - a review of recent developments. *BioResources*. https://doi.org/10.15376/biores.7.2.2506-2552.

Lawal, A. Q. T., Ninsiima, E., Odebiyib, O. S., Hassan, A. S., Oyagbola, I. A., Onu, P., & Yusuf, A. D. (2019). Effect of unburnt rice husk on the properties of concrete. *Procedia Manufacturing, 35*, 635–640. https://doi.org/10.1016/j.promfg.2019.06.006.

Li, X., Jia, P., & Wang, T. (2016). Furfural: A promising platform compound for sustainable production of C4 and C5 chemicals. *ACS Catalysis*. https://doi.org/10.1021/acscatal.6b01838.

Liu, F., Ma, C., Gao, Y., & McClements, D. J. (2017). Food-grade covalent complexes and their application as nutraceutical delivery systems: A review. *Comprehensive Reviews in Food Science and Food Safety*. https://doi.org/10.1111/1541-4337.12229.

Lü, F., Ye, X., & Liu, D. (2009). Review of antimicrobial food packaging. *Nongye Jixie Xuebao/Transactions of the Chinese Society of Agricultural Machinery*.

Mao, C., Feng, Y., Wang, X., & Ren, G. (2015). Review on research achievements of biogas from anaerobic digestion. *Renewable and Sustainable Energy Reviews*. https://doi.org/10.1016/j.rser.2015.02.032.

Mao, L., Zhang, L., Gao, N., & Li, A. (2012). $FeCl_3$ and acetic acid co-catalyzed hydrolysis of corncob for improving furfural production and lignin removal from residue. *Bioresource Technology*. https://doi.org/10.1016/j.biortech.2012.07.058.

Mariscal, R., Maireles-Torres, P., Ojeda, M., Sádaba, I., & López Granados, M. (2016). Furfural: A renewable and versatile platform molecule for the synthesis of chemicals and fuels. *Energy and Environmental Science*. https://doi.org/10.1039/c5ee02666k.

McClements, D. J., Bai, L., & Chung, C. (2017). Recent Advances in the utilization of natural emulsifiers to form and stabilize emulsions. *Annual Review of Food Science and Technology*. https://doi.org/10.1146/annurev-food-030216-030154.

McCrone, A., Moslener, U., D'Estais, F., & Grünig, C. (2017). *Global trends in renewable energy investment 2017*. Frankfurt School UNEP Collaborating Centre for Climate and Sustainable Energy Finance.

Mohd, N., & Aziz, N. (2016). Performance and robustness evaluation of nonlinear autoregressive with exogenous input model predictive control in controlling industrial fermentation process. *Journal of Cleaner Production*. https://doi.org/10.1016/j.jclepro.2016.06.191.

Molenveld, K., van den Oever, M., & Bos, H. (2015). Biobased packaging catalogue. *Wageningen*.

Morais, C., Freitas, F., Cruz, M. V., Paiva, A., Dionísio, M., & Reis, M. A. M. (2014). Conversion of fat-containing waste from the margarine manufacturing process into bacterial polyhydroxyalkanoates. *International Journal of Biological Macromolecules*. https://doi.org/10.1016/j.ijbiomac.2014.04.044.

Munoz, L. E. A., & Riley, M. R. (2008). Utilization of cellulosic waste from tequila bagasse and production of polyhydroxyalkanoate (pha) bioplastics by Saccharophagus degradans. *Biotechnology and Bioengineering*. https://doi.org/10.1002/bit.21854.

Nunes, L. J. R., Matias, J. C. O., & Catalão, J. P. S. (2014). A review on torrefied biomass pellets as a sustainable alternative to coal in power generation. *Renewable and Sustainable Energy Reviews*. https://doi.org/10.1016/j.rser.2014.07.181.

Ocak, B. (2019). Development of the mechanical and barrier properties of collagen hydrolysate/carboxymethyl cellulose films by using SiO_2 nanoparticles. *Pamukkale University Journal of Engineering Sciences*. https://doi.org/10.5505/pajes.2018.80688.

Okcu, Z., Yavuz, Y., & Kerse, S. (2018). Edible film and coating applications in fruits and vegetables. *Alınteri Zirai Bilimler Dergisi*. https://doi.org/10.28955/alinterizbd.368362.

Onu, P., Abolarin, M. S., & Anafi, F. O. (2017). Assessment of effect of rice husk ash on burnt properties of badeggi clay. *International Journal of Advanced Research, 5*(5), 240–247. https://doi.org/10.21474/IJAR01/4103.

Onu, P., & Mbohwa, C. (2018a). Future energy systems and sustainable emission control: Africa in perspective. *Proceedings of the International Conference on Industrial Engineering and Operations Management.*

Onu, P., & Mbohwa, C. (2018b). Green supply chain management and sustainable industrial practices: Bridging the gap. *Proceedings of the International Conference on Industrial Engineering and Operations Management*, 786–792 (Washington DC).

Onu, P., & Mbohwa, C. (2019a). Sustainable production: New thinking for SMEs. *Journal of Physics: Conference Series, 1378*(2). https://doi.org/10.1088/1742-6596/1378/2/022072.

Onu, P., & Mbohwa, C. (2019b). Sustainable supply chain management: Impact of practice on manufacturing and industry development. *Journal of Physics: Conference Series, 1378*(2). https://doi.org/10.1088/1742-6596/1378/2/022073.

Onu, P., & Mbohwa, C. (2019c). Renewable energy technologies in brief. *International Journal of Scientific & Technology Research, 8*(10), 1283–1289.

Onu, P., & Mbohwa, C. (2019d). Industrial energy conservation initiative and prospect for sustainable manufacturing. *Procedia Manufacturing, 35*, 546–551. https://doi.org/10.1016/j.promfg.2019.05.077.

Paudel, S. R., Banjara, S. P., Choi, O. K., Park, K. Y., Kim, Y. M., & Lee, J. W. (2017). Pretreatment of agricultural biomass for anaerobic digestion: Current state and challenges. *Bioresource Technology*. https://doi.org/10.1016/j.biortech.2017.08.182.

Payne, R. (2009). Carbohydrates: The essential molecules of life by Robert Stick and Spencer J. Williams. *Australian Journal of Chemistry*. https://doi.org/10.1071/ch09202_br.

Pourzolfaghar, H., Abnisa, F., Daud, W. M. A. W., & Aroua, M. K. (2016). A review of the enzymatic hydroesterification process for biodiesel production. *Renewable and Sustainable Energy Reviews*. https://doi.org/10.1016/j.rser.2016.03.048.

Raji, I. O., & Onu, P. (2017). Untapped wealth potential in fruit for Uganda community. *International Journal of Advanced Academic Research, 3*(February), 17–25. Retrieved from http://www.ijaar.org/articles/Volume3-Number1/Sciences-Technology-Engineering/ijaar-ste-v3n1-jan17-p7.pdf.

Rao, U., Sridhar, R., & Sehgal, P. K. (2010). Biosynthesis and biocompatibility of poly(3-hydroxybutyrate-co-4-hydroxybutyrate) produced by *Cupriavidus necator* from spent palm oil. *Biochemical Engineering Journal*. https://doi.org/10.1016/j.bej.2009.11.005.

Roy Chowdhury, S., Kumar Basak, R., Sen, R., & Adhikari, B. (2011). Production of extracellular polysaccharide by *Bacillus megaterium* RB-05 using jute as substrate. *Bioresource Technology*. https://doi.org/10.1016/j.biortech.2011.03.099.

Sawant, S. S., Salunke, B. K., & Kim, B. S. (2015). Degradation of corn stover by fungal cellulase cocktail for production of polyhydroxyalkanoates by moderate halophile Paracoccus sp. LL1. *Bioresource Technology*. https://doi.org/10.1016/j.biortech.2015.07.019.

Seesuriyachan, P., Techapun, C., Shinkawa, H., & Sasaki, K. (2010). Solid state fermentation for extracellular polysaccharide production by lactobacillus confusus with coconut water and sugar cane juice as renewable wastes. *Bioscience, Biotechnology and Biochemistry*. https://doi.org/10.1271/bbb.90663.

Sharif, H. R., Williams, P. A., Sharif, M. K., Abbas, S., Majeed, H., Masamba, K. G., … Zhong, F. (2018). Current progress in the utilization of native and modified legume proteins as emulsifiers and encapsulants – a review. *Food Hydrocolloids*. https://doi.org/10.1016/j.foodhyd.2017.01.002.

Sharma, P., & Bajaj, B. K. (2015). Production of poly-β-hydroxybutyrate by Bacillus cereus PS 10 using biphasic-acid-pretreated rice straw. *International Journal of Biological Macromolecules*. https://doi.org/10.1016/j.ijbiomac.2015.05.049.

Shedrack, G. M., Yawas, D. S., Emmanuel, S. G., Peter, O., Dagwa, I. M., & Ibrahim, G. Z. (2019). Effect of postweld heat treatment on the mechanical behaviour of austinitic stainless steel. *Global Journal of Engineering Science and Research Management, 6*(4), 1–9. https://doi.org/10.5281/zenodo.2639252.

Sindhu, R., Silviya, N., Binod, P., & Pandey, A. (2013). Pentose-rich hydrolysate from acid pretreated rice straw as a carbon source for the production of poly-3-hydroxybutyrate. *Biochemical Engineering Journal*. https://doi.org/10.1016/j.bej.2012.12.015.

Sutanto, S., Go, A. W., Chen, K. H., Nguyen, P. L. T., Ismadji, S., & Ju, Y. H. (2017). Release of sugar by acid hydrolysis from rice bran for single cell oil production and subsequent in-situ transesterification for biodiesel preparation. *Fuel Processing Technology*. https://doi.org/10.1016/j.fuproc.2017.07.014.

Tabasum, S., Noreen, A., Maqsood, M. F., Umar, H., Akram, N., Nazli, Z.i. H., ... Zia, K. M. (2018). A review on versatile applications of blends and composites of pullulan with natural and synthetic polymers. *International Journal of Biological Macromolecules*. https://doi.org/10.1016/j.ijbiomac.2018.07.154.

Taherzadeh, M. J., & Karimi, K. (2008). Pretreatment of lignocellulosic wastes to improve ethanol and biogas production: A review. *International Journal of Molecular Sciences*. https://doi.org/10.3390/ijms9091621.

Taskin, M., Ozkan, B., Atici, O., & Aydogan, M. N. (2012). Utilization of chicken feather hydrolysate as a novel fermentation substrate for production of exopolysaccharide and mycelial biomass from edible mushroom *Morchella esculenta*. *International Journal of Food Sciences and Nutrition*. https://doi.org/10.3109/09637486.2011.640309.

Taylor, M. J., Durndell, L. J., Isaacs, M. A., Parlett, C. M. A., Wilson, K., Lee, A. F., & Kyriakou, G. (2016). Highly selective hydrogenation of furfural over supported Pt nanoparticles under mild conditions. *Applied Catalysis B: Environmental*. https://doi.org/10.1016/j.apcatb.2015.07.006.

Technology roadmap: Bioenergy for heat and power. *Management of Environmental Quality: An International Journal.* , (2012)https://doi.org/10.1108/meq.2013.08324aaa.005.

Tomorrowmachine.se. (2013). Retrieved August 22, 2019, from http://tomorrowmachine.se/.

Tortajada, M., da Silva, L. F., & Prieto, M. A. (2013). Second-generation functionalized mediumchain- length polyhydroxyalkanoates: The gateway to high-value bioplastic applications. *International Microbiology*. https://doi.org/10.2436/20.1501.01.175.

Trinetta, V., & Cutter, C. N. (2016). Chapter 30 – Pullulan: A suitable biopolymer for antimicrobial food packaging applications A2 – Barros-Velázquez, Jorge. In *Antimicrobial food packaging*. https://doi.org/10.1016/B978-0-12-800723-5.00030-9.

Urbaniec, K., & Bakker, R. R. (2015). Biomass residues as raw material for dark hydrogen fermentation – a review. *International Journal of Hydrogen Energy*. https://doi.org/10.1016/j.ijhydene.2015.01.073.

Val del Río, Á., Campos Gómez, J. L., & Mosquera Corral, A. (2016). *Technologies for the treatment and recovery of nutrients from industrial wastewater*. https://doi.org/10.4018/978-1-5225-1037-6.

Verlinden, R. A. J., Hill, D. J., Kenward, M. A., Williams, C. D., Piotrowska-Seget, Z., & Radecka, I. K. (2011). *Production of polyhydroxyalkanoates from waste frying oil by cupriavidus necator*. AMB Express. https://doi.org/10.1186/2191-0855-1-11.

Vroman, I., & Tighzert, L. (2009). Biodegradable polymers. *Materials*. https://doi.org/10.3390/ma2020307.

Xiang, Y., Xu, X., & Li, J. (2012). Chemical properties and antioxidant activity of exopolysaccharides fractions from mycelial culture of Inonotus obliquus in a ground corn stover medium. *Food Chemistry*. https://doi.org/10.1016/j.foodchem.2012.03.121.

Yan, K., Wu, G., Lafleur, T., & Jarvis, C. (2014). Production, properties and catalytic hydrogenation of furfural to fuel additives and value-added chemicals. *Renewable and Sustainable Energy Reviews*. https://doi.org/10.1016/j.rser.2014.07.003.

Yusuf, A. A., Peter, O., Hassanb, A. S., A, L. T., Oyagbola, I. A., Mustafad, M. M., & Yusuf, D. A. (2019). Municipality solid waste management system for Mukono district. *Procedia Manufacturing, 35*, 613–622. https://doi.org/10.1016/j.promfg.2019.06.003.

Zhang, T., Li, W., Xu, Z., Liu, Q., Ma, Q., Jameel, H., … Ma, L. (2016). Catalytic conversion of xylose and corn stalk into furfural over carbon solid acid catalyst in γ-valerolactone. *Bioresource Technology*. https://doi.org/10.1016/j.biortech.2016.02.108.

Zhang, J., Zhang, B., Zhang, J., Lin, L., Liu, S., & Ouyang, P. (2010). Effect of phosphoric acid pretreatment on enzymatic hydrolysis of microcrystalline cellulose. *Biotechnology Advances*. https://doi.org/10.1016/j.biotechadv.2010.05.010.

CHAPTER

5 Sustainable agrowaste diversity versus sustainable development goals

5.1 Introduction

The global increase in energy demand caused by the growing world population figure is having a drastic effect on resource management, especially in developing economies, compared with advanced economies across the world. The capacity to effectively manage agricultural resources and embrace emerging technoeconomic innovations to maintain the high food demand is contingent on the implementation of nascent strategies and sustainable models (Onu & Mbohwa, 2019e). Hence, the evolution and development of ideas and expansion of interest based on global perspective inspired the September 2000 Millennium Summit with the 189 countries in attendance that lead to the first set of the millennial goals. Subsequently, the gathering of United Nations (UN) to address the pertinent issue concerning humanity and energy on sustainable development unanimously declares 2014–2024 as "Decade of Sustainable Energy for All" (United Nations, 2014), making it the 21ST agenda, specifically to "ensure access to reliable, affordable, and sustainable modern energy."

Since then, the United Nations assembly which held on September 25, 2015 in New York, met by a conscientious agreement on the issue of sustainability with a target to transform the current global socioeconomic and environmental experiences by 2030, penned down as the 17 Sustainable Development Goals (SDGs) (Siew, 2015). The 2015 meeting featured more countries (195), and the goal was to deliberate how the 5-P focus areas: "People," "Planet," "Prosperity," "Peace," and "Partnership" can align with the objectives of the 17 SDGs. This is illustrated as seen in Fig. 5.1. Meanwhile, much emphasis is centered on energy shortages and environmental abasement problems and collaborations to develop alternative technologies for future gains. According to Kumar et al. (2017), the 2015 agreement has led to a more conscious search for renewable, sustainable, and low energy-intensive technologies.

Agricultural Waste Diversity and Sustainability Issues. https://doi.org/10.1016/B978-0-323-85402-3.00012-7

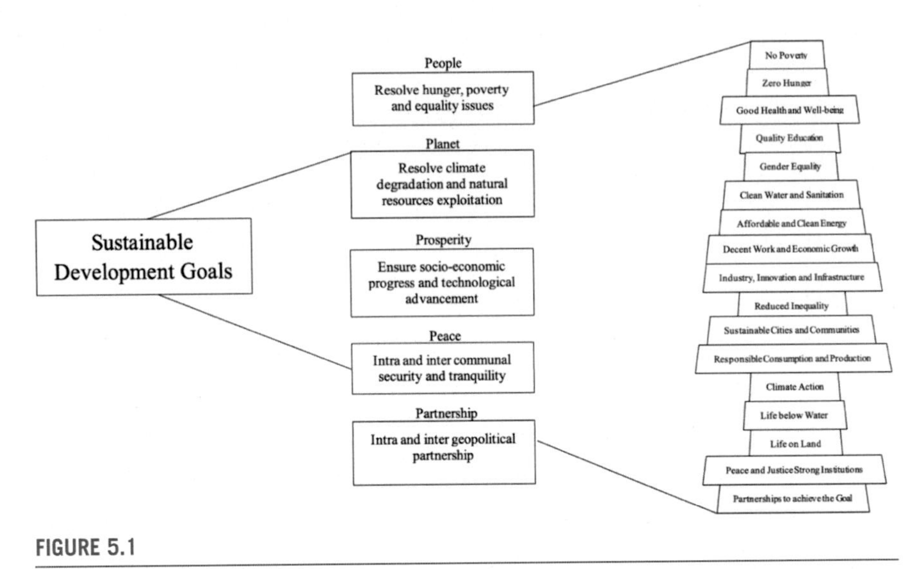

FIGURE 5.1

5-P perspective to achieving sustainable development goal.

The sustainable development agenda and the interplay of biodiversity

The concept of sustainable development has been primarily discussed and adopts an acceptable definition (Baumgartner & Rauter, 2017; Siew, 2015), which is first published by the UN commission, according to the Brundtland report of 1987 thus: "development that meets the needs of the present generation without compromising the ability of future generations to meet their own needs" (Brundtland, 1987). The report draws particular interest to three pillars: economic growth, environmental conservation, and social equality (Akiyode, Tumushabe, Hadijjah, & Onu, 2017; Onu & Mbohwa, 2019a). With regard to diversifying agro-based food wastes, the unwanted materials originating from food processing, including meat and fish, which are in enormous supply at the animal slaughterhouses, and processing plants serve as essential resources (Jayakumar, Selvamurugan, Nair, Tokura, & Tamura, 2008). Also, fruits and vegetable processing, including cash crops (wastes from coffee, tobacco, and cocoa farming, etc.), among others, serve the sugar factories and dairies manufacturers (Bosco & Chiampo, 2010), all serve as resource centers for biodiversification opportunities. Discards from the production of baked foods and sweets (Leung, Cheung, Zhang, Lam, & Lin, 2012), or from the brewing process of alcoholic and nonalcoholic beverage production, contribute to the waste from the overall agricultural process (Moralejo-Gárate et al., 2014). Moreover, the agricultural, food, and beverage processing industries have been reported to fall under the group of substantial energy consumer (Lin & Xie, 2015; Xu & Szmerekovsky, 2017); they are also significant contributors of unwanted wastes (Waldron, 2009). As such, they necessitate the application and adoption of the most advanced strategies and methodologies to ensure sustainability (Brunner & Rechberger, 2015; Chabbi et al., 2017; Zaman, 2015).

5.2 Achieving sustainable development goals versus agricultural waste diversity

SDG 1: To end poverty in all its forms everywhere

Modernized farming and other agricultural operations are cited on the outskirt or mostly conducted in isolated islands, and communities located far away from the township (domestic/financial settlements) due to the space consideration. The odds support job creation in the area and attract opportunities away from the cities. Hence, the depopulation of the urban areas while reducing pressure on resources, in the long run, increases income and reduces poverty (de Janvry & Sadoulet, 2009; Wang, 2013). More so, the expansion of the agroindustries will allow the sector to produce more food, against the fear of the rising number of global population index, as well as attract dividend from biofuels production. This indirectly creates opportunities for new corporations, and SMEs to spring up, leading to the creation of more jobs (Hawkes & Popkin, 2015; Pogge & Sengupta, 2016; Tewdwr-Jones, 2015). The outcome will be the result of more registered taxpayers, poverty reduction, and income to fund other government projects (Katikireddi & Dundas, 2017; Onu & Mbohwa, 2018a; Onu & Mbohwa, 2020).

In a recent study involving farm operations, the data show that households closest to modernized farming areas enjoyed a better standard of living, compared with far away dwellers due to the presence of dynamic economic activities (AGRA, 2017; Sheahan & Barrett, 2017). A well-organized farm would apply the services of administrative employees, machine operators, cleaners, and transportation operatives, among others, to effectively carry out their activities. Unfortunately, the situation in Africa today, and the sub-Saharan Africa context, sees laborers suffer from lack of favorable local content policy and, thus, the need to review industry/farm workers entitlement and work conditions. This is indeed rising and pressing concern in the 21st century, as labor is exploited and the majority of the workforce available in farms of this nature is conditioned to unfair treatments (Onu & Mbohwa, 2019b; Pogge & Sengupta, 2015). Achieving the current SDG 1 (Kenny, 2015), from the perspective of agricultural development to eradicate hunger, must include strategies that prioritize local content requirement to improve livelihood in that communities where they are cited, and the geopolitical region at large.

SDG 2: To end hunger, achieve food security and improved nutrition, and promote sustainable agriculture

The relationship between sustainable agricultural waste diversity, and effective agriculture operations, with soil and water maximization, to achieve food security is juxtaposed in this section. The most conventional practice of citing a commercial farming project is to locate it in a rural settlement, where sustainable interventions can be implemented. As such, the requirement for the actualization of the SDG 2 (Lartey, 2015) is to conservatively carry out farming operations and utilize waste resources from agriculture, while maximizing space, and without contaminating the ecosystem around that location (living beings, air, water, and soil). Incessant dumping of agricultural wastes may result in adverse challenges that are highly costly, detrimental to soil contamination, water toxification, and damage to land, compromising food quality and environmental insecurity in the region (Ringler et al., 2016).

The need to contribute to sustainable agricultural advancement necessitates exploitation and management of the resources and systems that sustain the process to meet adequate food supply for quality nourishment to eradicate hunger (Charlton, 2016; Melorose, Perroy, & Careas, 2015). Inspection and monitoring of agricultural activities and the causative effects on the produce, or impact of the systems, and cost implication of the agrowaste diversity protocol will assure food security and palliate the level of hunger. Other control strategies that can improve food quality include bioremediation practices to maintain resources (soil and water condition), and the application of different approaches, that can promote ecosystem conservation (MacKinnon, Dudley, & Sandwith, 2011).

SDG 3: To ensure healthy lives and promote well-being for all at all ages

Agricultural activities, and how it impacts the communities closest to the farms, differ from one geopollical location to another and impact on the health and well-being of settlers in that area. To implement change, ameliorate the tendency that affects healthy living, and promote well-being, the operational processes and work conditions, which guarantee the safety of lives and properties to ensure protection against accidents and diseases resulting from the activities of farming and wastes handling, must be reviewed and consistently improved (Singh Sekhon, 2014; Vuitton & Dalphin, 2017). Ensuring access to information concerning harmful chemicals and the physical presence of wastes that causes unwell to humans is advised. Awareness programs that will teach residence in the nearby community are necessary by sensitizing the population about the activities of the farming operation and the sustainable contribution put in place to ensure risk avoidance and the wealth and well-being of the community. Sustainability advancement and development, brought about by the integration of waste management programs and practices, will lead to new profit-making dimensions, leading to job creation and income generation (Lal, 2016; Tanaka, 2011). This will also promote benefits to workers and enrich their immediate households to meet necessary and essential needs, which will promote healthy living and well-being.

SDG 4: To ensure inclusive and equitable quality education and promote lifelong learning opportunities for all

An assessment of the connection between energy generation from agricultural wastes and education has implied few perspectives by researchers (Duncan & Ricketts, 2008; McKay, 2018; Soliva et al., 2007; Tang, 2015). The present goal is targeted at increasing literacy to forester substantial information dissemination and knowledge of the most relevant concepts of sustainable development. It focuses on the young and old, without gender bias. The United Nations charges the UNESCO to "provide global and regional leadership in education and responds to contemporary global challenges through education with a special focus on gender equality and Africa" (United Nations Educational Scientific and Cultural Organization, 2013). The idea is to make education and training on the subject matter; the potentials to gainful opportunities to all thus must become part of a lifelong learning and developmental framework. The agroindustrial sector must evolve in their operations and resources handling, or waste management to increase output (Owens, 2017). Since biodiversity has become a household name, and waste management practice requires innovation, it accentuates the need to facilitate sponsorship and inclusion of the field of study in higher education institution curriculum. Funding support purposefully for community dwellers and children of employees who work in the areas of agrowaste beneficiation (building of training centers, technical colleges, and universities) should be meant to foster inclusivity and sustenance of the opportunities that it offers to people.

Organizing industrial visits for a young scientists and distant learning programs about agricultural development and waste to wealth practices will position young learners to innovate future solutions for entrepreneurs to contribute in the actualization of the SDG (Borthakur & Govind, 2017; UNESCO, 2016). Information should be made clear and precise, and the educational material should be transcribed to the community's language where the interest for training is established in rural settlement (Moh & Abd Manaf, 2014; Yogo, 2017). In this view, lifelong learning opportunities for inclusive and equitable quality education can be met.

SDG 5: To achieve gender equality and empower all women and girls

The SDG 5 shows no direct correlation with sustainable agrowaste diversity; however, the collective activities in agricultural development and biodiversity offer opportunities have been mostly considered to fit the masculine job description. This bias reduces the working capacity and labor potential to achieve operational efficiency, as the efforts of all gender in agrowaste resources management may count for something. The processes of harvesting energy from sources or conversion of the agricultural waste to valued product interlink the role of women to impact positively on the actualization of the SDG 2030 targets (Akter et al., 2017; Huyer, 2016). All gender is responsible for the actualization of the sustainable development goals, without discrimination of any kind. So far, the percentage of women in different companies is rising steadily as the campaign for women liberation and gender equality increases globally.

For example, the South Africa mining sector maintains (hires) up to 20% female workers, compared with the United Kingdom with only less than 5% (Macedo, Boava, Cappelle, & Oliveira, 2012; PWC, 2013). Regardless, women's participation in agriculture development is less than 10% in the world. However, this value is higher than the global average in the sub-Saharan African region (up to 20%) (Kelsey, 2014). The figures reflect the inequality based on gender, not to mention the disparity that exists among enterprises where female workers receive low income. Hence, new waste management, biodiversity project, and existing agricultural programs must consider the women's and girls' input and gender equality for the wholistic contribution to economic, social, and environmental improvement and collective actualization of the SDGs.

SDG 6: To ensure availability and sustainable management of water and sanitation for all

Sustainable development and the aspect of waste management and bioenergy production cannot be separated from the challenges of adequate, quality, and viable water supply, hence ensuring water treatment/recycling, after the different processes that include preparation and production (Masi, Rizzo, & Regelsberger, 2018; Yusuf et al., 2019). Also, great concern for the health and safety of the surroundings is essential to monitor sanitation and quality of the water being used, whether infected from natural activities or due to the industrial events linked to release of harmful liquid and gaseous pollutants (Arhin, Boansi, & Zango, 2016; Brookes & Carey, 2015; McInnes, 2018). The course of sustainable development and industrialization will require firms to periodically submit reports on their waste management protocols and conservational tactics or ecodiversity initiatives (Onu & Mbohwa, 2019f). Although most biodegradable wastes are recoverable, and can be converted or reused in a sustainable manner, very few African countries are footing this front and continuing to diligently finance projects on the subject matter.

More so, there are no clear indices of any African nation, involved in exploiting the paradigm to meet energy needs (Nhamo, Nhemachena, & Nhamo, 2019). The recommendation for Africa and the sub-Saharan African region is to commit more attention to the value chain of water supply and conservation. Government and community involvement in the promotion of sustainable management, regarding water and sanitation for all must project the necessity through funding, developmental policy on water usage, and establishment of waste generation restrictions to alleviate environmental pollution while applying the different tactics elucidated by McInnes (2018).

SDG 7: To ensure access to affordable, reliable, sustainable, and modern energy for all

Here, cleaner energy production initiatives are brought to bear on actualization of the sustainable development goals, to ensure the consideration of the different forms, sources, and nature of technology required for affordable, reliable, sustainable, and modern energy development (Choi, Cao, & Zhao, 2016). Recently, industrial food manufacturers are focusing on improving their energy efficiency, thus ensuring optimal energy usage during production. This is as a result of effective operability while ensuring maintenances of the energy infrastructure are routinely conducted. The adoption of sustainable energy resources (cells) as alternatives to the traditional fossil utilization and the use of wind, solar, geothermal, or safe-to-operate nuclear energy systems are gaining prominence from the global perspective (United Nations, 2015, 2018). However, the contribution of the agroindustrial sector provides clean, green fuel among other alternative resources such as biomass to meet energy distribution.

The state of affairs of countries that have hugely invested in renewable energy technologies, in comparison with their CO_2 emission index, presents Brazil with the most commendable energy production potential status. While other renewable energy investor and tech giants such as China and the United States registered CO_2 emissions index of 772.1 kg/MWh and 489.8 kg/MWh, Brazil had only about 101.3 kg CO_2/MWh with a share of up to 43% renewables in the country's energy mix (EPE, 2016). Recently, the concern of rising population figure has been attributed to have the geometric increase in agricultural food farming and processing activities, which can directly impart energy utilization, therefore, challenges of electricity distributions. The various stages in waste management operations, prevention, preparation recycling/processing, recovery/conversion, and disposal apply different consumption options and energy strategies (Gavrilescu, 2008; Pickin & Randell, 2017). These stages need to be evaluated for their specific energy consumption based on the capacity and the magnitude of agricultural operations and engagement, or the different equipment used in the preparation (separation and treatment), transportation, and heating/cooling, to packaging and up to the final stage of disposal (Girotto, Alibardi, & Cossu, 2015; Hakami, 2016; Madurwar, Ralegaonkar, & Mandavgane, 2013).

SDG 8: To promote sustained, inclusive and sustainable economic growth, full and productive employment, and decent work for all

Progresses in new technology development for agricultural/farming expansion are becoming the main attraction in recent times. The relationship between employment creation to support the labor force and encourage the participation of the young working for meaningful contribution to a sustainable economy defines the SDG 8 (Parisotto, 2015; Rai, Brown, & Ruwanpura, 2019). The agroindustrial sector, through the various waste management schemes and energy generation strategies, has a chain of employment opportunities and income generation that can transform economies and improve livelihood (Frey, 2017; Onu & Mbohwa, 2018b). The need to control risky activities and conditions that hinder organizational development or the role of practitioners and workers to improve productivity in waste management for energy exploitation from agricultural and food processing operation is long overdue. The emphasis must be on Africa's strategy and the measures that can be applied to create new opportunities, significant to proffer solutions that will address the ailing employment situation. Regardless, already existing companies must also consider the necessity to provide personal safety protective equipment, avoid accident potentials engagements, and monitor/prevent unwanted incidence, diseases, or any form of catastrophic implication (Khan, Mustaq, & Tabassum, 2014). "Workers union" or "labor force societies" together with the government are enjoyed to create the guiding principles for operations and policies regarding employee engagement/remunerations while on the job (Sousa, Almeida, & Dias, 2014; Tchiehe & Gauthier, 2017).

SDG 9: To build resilient infrastructure, promote inclusive and sustainable industrialization, and foster innovation

The emergence of new techniques and development of nascent systems and management strategy linked to production operation contributes to the several opportunities originating from the agroindustrial sector, thus corroborating SDG (United Nations, 2016a). Recently, the campaign for indigenous products and raw materials development for high priority projects with locally sourced materials and skill sets has gained global support and practice. Hence, it ensures inclusivity (Mkumbo, 2016; UNIDO, 2017). The most recent initiatives in technological advancements in agriculture point to the industry 4.0 paradigm and role in agrowaste management (Dalenogare, Benitez, Ayala, & Frank, 2018; Müller, Jaeger, & Hanewinkel, 2019; Müller, Kiel, & Voigt, 2018). Digitization of processes and the use of robots have become a new trend in the general industrial space (UNIDO, 2017). Hence, it requires the commitment of governmental and private partners or investors to collaborate on the proliferation of resilient hardware and deployment of scientific discoveries, and research breakthroughs to promote an innovative solution.

SDG 10: To reduce inequality within and among countries

One of the purposes of SDGs that directly considers indigenous empowerment is to ensure equality (Hosseinpoor, Bergen, & Magar, 2015; United Nations Development Program, 2016). This provides the opportunity for "experts" and potential beneficiaries in whose corridors industrial advancement strives. More so, the issues over extreme discrepancies in wages payable to local employees compared with "expatriates" require a changed mindset and government support to ensure fairness to all involved (Chancel, Hough, & Voituriez, 2018). Organizing technical training and workshops for new and existing employees without bias or marginalization of any kind, and ensuring equitable allocation of resources between the international and indigenous investors in agricultural development, will foster sustainable development. The community must be carried along in the waste to wealth business process (Hackl, 2018; MacNaughton, 2017). The proposal ought to be viewed through the lens of equitability and midlevel/vocational programs that can be organized to educate capable individuals to explore the opportunities. More so, the promotion of equality will require supportive legislative injunctions that harmonize SDGs 4, 5, 8, and 9 to succeed regarding agroindustrial biomasses conversion for energy and materials production (Bhagat, 2017; Oestreich, 2018; Zimm, 2019).

Communities that have potentials for sustainable bioresources or are economically viable for exploitation must strategize with their government to facilitate projects actualization in line with labor, safety, and licensing consideration. This will help to ameliorate inequality and promote the present sustainable development goal.

SDG 11: To make cities and human settlements inclusive, safe, resilient, and sustainable

This SDG is compared with SDG 5, which does not directly affect sustainable agro-waste diversity. However, the desire to promote growth and development of social and environmentally sustainable projects in cities and local communities is linked to the evaluation of their impact on livelihood. The laws and policies that support urbanization also consider the possible challenges and consequences, as to the crises emanating from communities grieved due to land and river pollution, leading to a threat and violent action against workers and representatives in charge of the related projects (Daniel, 2015; Klopp & Petretta, 2017). This later results in community disputes and displacement of small businesses (Caprotti et al., 2017; Jung & Song, 2018). While the need for resilient urban settlement is pivotal for achieving the SDG, the justification for human satisfaction and inclusiveness must not be disregarded. Thus, the implementation of agrowaste conversion strategy that imbibes safe and conservative methods should be exploited. More so, a consensus between community representatives and government intervention must devise a balance and present a clear strategy about promoting technology advancement and community development.

SDG 12: To ensure sustainable consumption and production patterns

Agrowaste diversity and the principle of conservativeness to meet positive social, economic, and environmental gains create values from potential discards for economic growth and ecosystem preservation. Experts and practitioners in the agroindustrial sector are saddled with the responsibility to operate effectively, while improving waste management throughout, by adopting the Six-R (reduce, recover, reuse, redesign, recycle, remanufacture) mechanism, to achieve sustainable production and consumption patterns (Onu & Mbohwa, 2019f). In developing countries, new laws to access agrowastes from agricultural materials or the conditions for reuse are being established. Collection protocols and strategies that support robust waste management designs within farms and the food production and processing industries must consider space, energy, and environmental impact of the activities. To imbibe a circular economy and practices of reuse and waste management, both the material and energy option must be regularized in line with a sustainable disposition to satisfy the elements of the triple bottom line (social, economic, and environmental) (Vergragt, Akenji, & Dewick, 2014). The strategy will cause a change in the consumption and production pattern, promote waste reconversion, and drive the sustainable development goals (Hoballah & Averous, 2015).

SDG 13: To take urgent action to combat climate change and its impacts

The negative implication of the high volume of carbon released into the atmosphere over the past centuries, and actively, since the first industrial revolution era that has led to the current climatic situation, leaves the world in shocked. While industrialization continues to grow, the search for more lands is a threat to forestation and increases the impact of climate change. However, sustainable agrowaste management offers the potential for climate change mitigation, with the option to reduce methane emission from landfills and carbonaceous release from burning what is regarded to waste (Bruce et al., 2018; Rice, Burke, & Heynen, 2015). The call for a global reduction in carbon footprint must incorporate sustainable strategies, which are supported by government sanctions and laws against pollutive activities, leading to environmental contamination. Hence, achieving the current sustainable development goal with regard to reduction in ozone depletion, and global warming effects, will extend the issue of felling of trees, deforestation, forest fire, and agricultural exploitation, etc., to have a consolidated solution for conserving the natural ecosystem (Maupin, 2017). Also, despite the impact of the climate change which is detrimental to the survival of the living organism, the planned action to reduce or manage the activities that lead to its adverse effect becomes a means for social, economic, and environmental engagement to promote sustainability (Csa et al., 2014).

Funding of ecodiversity investigations and the institutions that will be responsible for monitoring and reporting, designing risk control measures, and technological interventions that will mitigate climate change effects will promote this SDG (Figueres, 2015). International collaborations between government and other industry stakeholders to support programs and commitment to tackle global climate change challenges is timely and will advance innovations in areas of agricultural waste diversity for materials and bioenergy production. In essence, it will serve as an essential and sustainable approach to drastically reduce methane emission and the dangerous effluents associated with fossil fuel use (Williams et al., 2015).

SDG 14: To conserve and sustainably use the oceans, seas, and marine resources for sustainable development

The use of resources that are available offshore has been explored (Cicin-Sain, 2015) and does not correlate with agricultural development through wastes conversion, in addition to SDGs 5 and 14. Notwithstanding, the 14th agenda depends on the success of the different interventions through which sustainability is achieved on land, considering most developing nations dump their agricultural wastes into large water bodies. Invariably, the oceans, seas, and marine resources are indirectly threatened by climate change variation (Recuero Virto, 2018) and humans. The Antarctica region have observed consistent rise in the sea level, while there is rapid dehydration and decrease in the Arctic region, thus affecting agriculture (farming operation) in the two regions. In addition, the role of human, due to incessant dumping of organic wastes and natural events such as chemical washoff into the rivers, leads to pollution and destruction of the ocean ecosystem (Mendler de Suarez, Cicin-Sain, Wowk, Payet, & Hoegh-Guldberg, 2014). Several construction activities, mining, and installation of renewable and alternative power systems offshore also affect life at sea, due to toxic and uninhabitable ecosystem (Onu & Mbohwa, 2018a).

The main reason for the SDG 14, which is closely linked to biodiversity, points to aquatic farming and the by-products from its industrial processing that has become of significant potential in pharmaceutical, aromatic, and food production companies, not to mention its wastewater potential for bioenergy production.

SDG 15: To protect, restore, and promote sustainable use of terrestrial ecosystems, sustainably manage the forest, combat desertification, halt and reverse land degradation, and halt biodiversity loss

The connection between sustainable agrowaste diversity and the concern over the balance in natural habitats: to protect the ecosystem, and enrich lands, from the perspectives of clean energy production, agricultural activities, that can contribute to the curtailment of desertification and land degradation, hence putting the lands to useful disposition (Schroeder, Anggraeni, & Weber, 2019). Exploration and recovery of areas for the advancement of renewable and natural resources promulgation must consider the need to avoid harmful waste dumpings and release dangerous chemicals (Onu & Mbohwa, 2018c), thus ensuring the need to commercialize the development of energy-based crops as a means to revive the terrestrial ecosystems. Hence, it will require governmental support and cooperation of practitioners involved in policies and funding to supports and promote biodiversity. A collaborative effort is also required through research, investment, monitoring, and project implementation involving different stakeholders to achieve this SDG (Mohieldin & Caballero, 2015).

SDG 16: To promote peaceful and inclusive societies for sustainable development, provide access to justice for all and build effective, accountable, and inclusive institutions at all levels

The goal here is to forester peace, justice, and tranquil, at all levels of the business and operational engagements to promote inclusive societies for sustainable development. This is important to bring about effectiveness, accountable, and progressive institutions at all tiers of development (Onyekwere, 2016; Revi & Rosenzweig, 2013). Adequate communication and information dissemination will increase awareness of workers, having the sense of satisfaction and partnership in the project (Attree & Möller-Loswick, 2015). Also, while the procedural commitment and protocols for project execution and services regarding waste management activates may involve fraud and misconduct, ethical integrity should not be overlooked (McClintock & Bell, 2013). While there is a need to promote local content laws and protect indigenous interests, more is desired as per transparency and disclosure on the business operations and deliverables (Biermann et al., 2009; Ivanovic, Cooper, & Nguyen, 2018).

SDG 17: To strengthen the means of implementation and revitalize the global partnership for sustainable development

This SDG focuses on international collaboration impartation, through exchanging ideas and standards for the improvement of operations, as per strategies that can be applied for energy production from agrowastes within the coverage of achieving sustainable growth (Seth, 2015). More so, the adherence to prevailing laws, operational guidance, and implementation capability can be used as a basis for performance comparison in the agroindustrial sector (Baker, 2015; Ford, 2015; Gusmão Caiado, Leal Filho, Quelhas, Luiz de Mattos Nascimento, & Ávila, 2018). Hence, the interlink between sustainable agricultural waste diversity and the need to foster global partnership for sustainable development must be committed to sharing breakthroughs on natural resources conservation, societal satisfaction, and supporting policies (Baumgartner & Rauter, 2017). In conclusion, achieving this last and final SDG has more to do with the nature of intergovernmental treaties and bilateral agreements if they exist.

5.3 Business as usual versus sustainable development goal implementation: biodiversity in sub-Saharan Africa

Sustainability within the context of clean energy production is transitioning through the use of efficient and effective technologies that generate reliable and nontoxic fuels from durable resources, ensuring social, economic, and environmental surety where people can experience better living standards and assurances of a habitable ecosystem. As part of the sustainability campaign to end poverty, promote climatic situation, and ensure profitability (United Nations, 2016b), industrial energy sectors are charged to meet secure, affordable, and assessable energy for all, through an environmentally friendly manner—Energy Trilemma, according to the World Energy Council (WEC, 2017). Liu, Liu, Yang, Chen, & Ulgiati, 2016 termed sustainability to be the most active component besides the uncertainty of renewable energy systems (RES) integration, to drive corporate strategies on environmental pollution reduction and promotion of climate mitigation (Liu, Liu, Yang, Chen, & Ulgiati, 2016). Hence, sustainability assessment becomes pivotal for a sustainable, eco-friendly society, providing information relevant to make adequate decisions concerning the future (Engert, Rauter, & Baumgartner, 2016).

Additionally, the deliberation on green initiatives, policies, and technologies that will align biodiversity with the sustainable development goals is pertinent, for the sake of resolving hunger, poverty, and inequality issues among the people; address climate degradation and overexploitation of the planet's natural resources; achieve prosperity through socioeconomic and technological progress; achieve peace through fostering of security; promote inclusiveness on a local and international basis; and finally, embrace partnership which has to do with collaboration and exchange of innovation/technology transfer in the sub-Saharan African region. This corroborates with the assertions of Blicharska et al. (2019) that biodiversity is not only relevant to SDGs

14 and 15 but is also linked to the actualization of 10 SDGs. This is demonstrated in Fig. 5.2, with the SDGs 1, 2, and 6 indirectly impacting at least four other SDGs concurrently (SDGs. 4, 5, 10, and 16) and SDG 17 by SDGs 1, 2, 3, 7, 11, and 13.

The goal of biodiversity, linked to environmental pollution resolution, can be achieved premise the following: level of commitment of the entities involved, community interest, management perspective of the waste treatment organization (mission/motivation/purpose of operation), and strategy for integration of the sustainable initiatives and green technology proliferation. The primary considerations for the sub-Sharan Africa are based on organizational culture and change implementation perception to advance sustainability as developmental paradigm. It affects the general motivation of the workers and should be approached cautiously with the

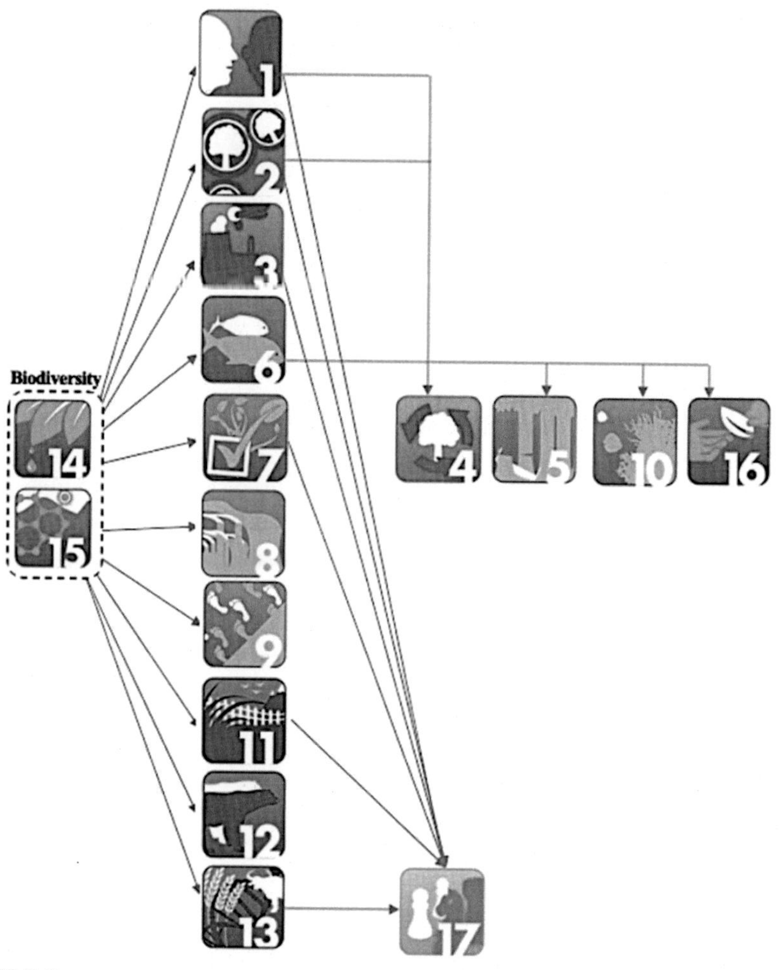

FIGURE 5.2

Schematic of biodiversity's contributions to sustainable development.

right strategy by the organizational development practitioners in charge to eliminate poverty; provide worldwide access to food, clean water, energy, health care, and education; achieve gender equality; guarantee a decent job for everyone; build sustainable infrastructures; reduce income inequality; promote urban development, rational consumption, and production; solve the problem of climate changes; save oceans; prevent deforestation; and form structures necessary to achieve these goals, including a global partnership that boosting sustainable development. While SDG 7, access to affordable, reliable, sustainable, and modern energy for everyone, stands out the most, SDGs 5, 11, and 14 least support the proposition of sustainability and agricultural waste diversity.

5.4 Conclusions and perspectives

Scientists continue in the study of how energy technologies influence sustainable development, in terms of their contribution to global environmental issues (Onu & Mbohwa, 2019c, 2019d). Lately, the influence of energy technologies on sustainable development actualization is being researched from the viewpoint of the industry 4.0 dimension to achieving the SDGs. Since the UN General Assembly, which led to the resolution "Transforming our world: the 2030 Agenda for Sustainable Development" in 2015, several nations of the sub-Saharan Africa countries have made remarkable progress in renewable energy advancement and considered biofuels from easily accessible biodegradable agricultural wastes for powering cities and public transport systems. This is evident in South Africa, Angola, Morocco, and Uganda, to mention a few. Hence, it requires the commitment of governmental, public and private stakeholders, or investors to collaborate on the proliferation of resilient scientific innovations and the application of research breakthroughs to promote a substantial solution to the challenges which the SDGs hope to address. As such, new policies and funding schemes that will promote biodiversity, education, monitoring, and project implementation involving the different techniques to achieve the SDG must become the priority in Africa.

References

AGRA. (2017). *Africa agriculture status report: The business of smallholder agriculture in sub-Saharan Africa*. AGRA. http://hdl.handle.net/10568/42343.

Akiyode, O., Tumushabe, A., Hadijjah, K., & Onu, P. (2017). Climate change, food security and environmental securi-ty: A conflict inclination assessment of Karamoja region of Uganda. *International Journal of Scientific World, 5*(2), 1–5. https://doi.org/10.14419/ijsw.v5i2.8458.

Akter, S., Rutsaert, P., Luis, J., Htwe, N. M., San, S. S., Raharjo, B., & Pustika, A. (2017). Women's empowerment and gender equity in agriculture: A different perspective from Southeast Asia. *Food Policy, 69*, 270–279. https://doi.org/10.1016/j.foodpol.2017.05.003.

Arhin, E., Boansi, A. O., & Zango, M. S. (2016). Trace elements distributions at Datoko-Shega artisanal mining site, Northern Ghana. *Environmental Geochemistry and Health, 37*(2), 203–218. https://doi.org/10.1007/s10653-015-9705-0.

Attree, L., & Möller-Loswick, A. (2015). Promote peaceful and inclusive societies for sustainable development, provide access to justice for all and build effective, accountable and inclusive institutions at all levels. *UN Chronicle*, 75–78. https://doi.org/10.18356/36c826cb-en.

Baker, S. (2015). *Sustainable development*. Routledge.

Baumgartner, R. J., & Rauter, R. (2017). Strategic perspectives of corporate sustainability management to develop a sustainable organization. *Journal of Cleaner Production*. https://doi.org/10.1016/j.jclepro.2016.04.146.

Bhagat, R. B. (2017). Migration and urban transition in India: Implications for development migration and urban transition in India: Implications for development. In *United nations expert group meeting on sustainable cities, human mobility and international migration*. https://doi.org/10.13140/RG.2.2.17888.17925.

Biermann, F., Betsill, M. M., Gupta, J., Kanie, N., Lebel, L., Liverman, D., … Siebenhüner, B. (2009). *Earth system governance: People, places and the planet*. Global Governance. https://doi.org/10.1787/9789264203419-101-en.

Blicharska, M., Smithers, R. J., Mikusiński, G., Rönnbäck, P., Harrison, P. A., Nilsson, M., & Sutherland, W. J. (2019). Biodiversity's contributions to sustainable development. *Nature Sustainability, 2*, 1083–1093. https://doi.org/10.1038/s41893-019-0417-9.

Borthakur, A., & Govind, M. (2017). Emerging trends in consumers' E-waste disposal behaviour and awareness: A worldwide overview with special focus on India. *Resources, Conservation and Recycling, 117*, 102–113. https://doi.org/10.1016/j.resconrec.2016.11.011.

Bosco, F., & Chiampo, F. (2010). Production of polyhydroxyalcanoates (PHAs) using milk whey and dairy wastewater activated sludge. Production of bioplastics using dairy residues. *Journal of Bioscience and Bioengineering*. https://doi.org/10.1016/j.jbiosc.2009.10.012.

Brookes, J. D., & Carey, C. C. (2015). Ensure availability and sustainable management of water and sanitation for all. *UN Chronicle*. https://doi.org/10.18356/d694f52d-en.

Bruce, M., C., James, H., Janie, R., Clare, M., S., Stephen, T., & Wollenberg, E. L. (2018). Urgent action to combat climate change and its impacts (SDG 13): Transforming agriculture and food systems. *Current Opinion in Environmental Sustainability, 34*, 13–20. https://doi.org/10.1016/j.cosust.2018.06.005.

Brundtland, G. H. (1987). *Our common future (Brundtland report)*. United Nations Commission.

Brunner, P. H., & Rechberger, H. (2015). Waste to energy - key element for sustainable waste management. *Waste Management, 37*, 3–13. https://doi.org/10.1016/j.wasman.2014.02.003.

Caprotti, F., Cowley, R., Datta, A., Broto, V. C., Gao, E., Georgeson, L., … Joss, S. (2017). The new urban agenda: Key opportunities and challenges for policy and practice. *Urban Research and Practice*. https://doi.org/10.1080/17535069.2016.1275618.

Chabbi, A., Lehmann, J., Ciais, P., Loescher, H. W., Cotrufo, M. F., Don, A., … Rumpel, C. (2017). Aligning agriculture and climate policy. *Nature Climate Change*. https://doi.org/10.1038/nclimate3286.

Chancel, L., Hough, A., & Voituriez, T. (2018). Reducing inequalities within countries: Assessing the potential of the sustainable development goals. *Global Policy*, 1–12. https://doi.org/10.1111/1758-5899.12511.

Charlton, K. E. (2016). Food security, food systems and food sovereignty in the 21st century: A new paradigm required to meet sustainable development goals. *Nutrition and Dietetics*. https://doi.org/10.1111/1747-0080.12264.

Choi, C. H., Cao, J., & Zhao, F. (2016). System dynamics modeling of indium material flows under wide deployment of clean energy technologies. *Resources, Conservation and Recycling, 114*, 59–71. https://doi.org/10.1016/j.resconrec.2016.04.012.

Cicin-Sain, B. (2015). Conserve and sustainably use the oceans, seas and marine resources for sustainable development. *UN Chronicle*. https://doi.org/10.18356/8fcfd5a1-en.

Csa, C., Farmers, S., Resilient, T., Livelihoods, S., Agriculture, C. S., Cline, W. R., … FAO. (2014). *Climate-smart agriculture: Managing ecosystems for sustainable livelihoods*. Cgspace.Cgiar.Org. https://doi.org/10.1186/s40066-016-0075-3.

Dalenogare, L. S., Benitez, G. B., Ayala, N. F., & Frank, A. G. (2018). The expected contribution of industry 4.0 technologies for industrial performance. *International Journal of Production Economics, 204*, 383–394. https://doi.org/10.1016/j.ijpe.2018.08.019.

Daniel, K. (2015). Make cities and human settlements inclusive, safe, resilient and sustainable. *UN Chronicle*. https://doi.org/10.18356/54bc8fad-en.

de Janvry, A., & Sadoulet, E. (2009). Agricultural growth and poverty reduction: Additional evidence. *World Bank Research Observer, 25*(1), 1–20. https://doi.org/10.1093/wbro/lkp015.

Duncan, D., & Ricketts, J. (2008). Total program efficacy: A comparison of traditionally and alternatively certified agriculture teachers. *Journal of Agricultural Education, 49*, 38–46. https://doi.org/10.5032/jae.2008.04038.

Engert, S., Rauter, R., & Baumgartner, R. J. (2016). Exploring the integration of corporate sustainability into strategic management: A literature review. *Journal of Cleaner Production, 112*(4), 2833–2850. https://doi.org/10.1016/j.jclepro.2015.08.031.

EPE. (2016). *Plano Decenal de Expansão - PDE 2026: Capítulo 9: Eficiência Energética e Geração Distriuída*. Plano Decenal de Energia.

Figueres, C. (2015). Take urgent action to combat climate change and its impacts. *UN Chronicle*. https://doi.org/10.18356/0ab994c7-en.

Ford, L. (2015). Sustainable development goals: All you need to know. *The Guardian*.

Frey, D. F. (2017). Economic growth, full employment and decent work: The means and ends in SDG 8. *International Journal of Human Rights, 21*(8), 1–21. https://doi.org/10.1080/13642987.2017.1348709.

Gavrilescu, D. (2008). Energy from biomass in pulp and paper mills. *Environmental Engineering and Management Journal*.

Girotto, F., Alibardi, L., & Cossu, R. (2015). Food waste generation and industrial uses: A review. *Waste Management, 45*, 32–41. https://doi.org/10.1016/j.wasman.2015.06.008.

Gusmão Caiado, R. G., Leal Filho, W., Quelhas, O. L. G., Luiz de Mattos Nascimento, D., & Ávila, L. V. (2018). A literature-based review on potentials and constraints in the implementation of the sustainable development goals. *Journal of Cleaner Production, 198*, 1276–1288. https://doi.org/10.1016/j.jclepro.2018.07.102.

Hackl, A. (2018). Mobility equity in a globalized world: Reducing inequalities in the sustainable development agenda. *World Development*. https://doi.org/10.1016/j.worlddev.2018.08.005.

Hakami, B. A. (2016). Waste disposal and landfill: Information needs. *International Journal of Civil Engineering and Technology*.

Hawkes, C., & Popkin, B. M. (2015). Can the sustainable development goals reduce the burden of nutrition-related non-communicable diseases without truly addressing major food system reforms? *BMC Medicine, 13*(1), 1–143. https://doi.org/10.1186/s12916-015-0383-7.

Hoballah, A., & Averous, S. (2015). Ensure sustainable consumption and production patterns. *UN Chronicle*. https://doi.org/10.18356/4bd7f86c-en.

Hosseinpoor, A. R., Bergen, N., & Magar, V. (2015). Monitoring inequality: An emerging priority for health post-2015. *Bulletin of the World Health Organization*. https://doi.org/10.2471/BLT.15.162081.

Huyer, S. (2016). *Closing the gender gap in agriculture*. (2nd ed., *20,* pp. 105–116). Gender, Technology and Development.

Ivanovic, A., Cooper, H., & Nguyen, A. M. (2018). *Institutionalisation of SDG 16: More a trickle than a cascade?* Social Alternatives.

Jayakumar, R., Selvamurugan, N., Nair, S. V., Tokura, S., & Tamura, H. (2008). Preparative methods of phosphorylated chitin and chitosan-an overview. *International Journal of Biological Macromolecules, 43*(3), 221–225. https://doi.org/10.1016/j.ijbiomac.2008.07.004.

Jung, T. Y., & Song, J. (2018). Goal 11: Sustainable city and communities. In C. T'ae-yong (Ed.), *Sustainable development goals in the republic of Korea*. UNDP.

Katikireddi, S. V., & Dundas, R. (2017). Relative poverty still matters. *The Lancet Public Health*. https://doi.org/10.1016/S2468-2667(17)30029-4.

Kelsey, N. (2014). *Revealing the gap between men and women farmers - national geographic*. Nat Geo Food. Retrieved September 13, 2019 from: https://www.nationalgeographic.com/news/2014/3/140308-international-female-farmers/.

Kenny, C. (2015). End poverty in all its forms everywhere. *UN Chronicle*. https://doi.org/10.18356/0aec43e1-en.

Khan, W. A., Mustaq, T., & Tabassum, A. (2014). Occupational health, safety and risk analysis. *International Journal of Science and Technology*.

Klopp, J. M., & Petretta, D. L. (2017). The urban sustainable development goal: Indicators, complexity and the politics of measuring cities. *Cities, 63*, 92–97. https://doi.org/10.1016/j.cities.2016.12.019.

Kumar, A., Sah, B., Singh, A. R., Deng, Y., He, X., Kumar, P., & Bansal, R. C. (2017). A review of multi criteria decision making (MCDM) towards sustainable renewable energy development. *Renewable and Sustainable Energy Reviews, 69*, 596–609. https://doi.org/10.1016/j.rser.2016.11.191. June 2016.

Lal, R. (2016). Environmental sustainability. *Climate Change and Multi-Dimensional Sustainability in African Agriculture: Climate Change and Sustainability in Agriculture*. https://doi.org/10.1007/978-3-319-41238-2_1.

Lartey, A. (2015). End hunger, achieve food security and improved nutrition and promote sustainable agriculture. *UN Chronicle*. https://doi.org/10.18356/5940d90a-en.

Leung, C. C. J., Cheung, A. S. Y., Zhang, A. Y. Z., Lam, K. F., & Lin, C. S. K. (2012). Utilisation of waste bread for fermentative succinic acid production. *Biochemical Engineering Journal, 65*, 10–15. https://doi.org/10.1016/j.bej.2012.03.010.

Lin, B., & Xie, X. (2015). Factor substitution and rebound effect in China's food industry. *Energy Conversion and Management, 105*, 20–29. https://doi.org/10.1016/j.enconman.2015.07.039.

Liu, X., Liu, G., Yang, Z., Chen, B., & Ulgiati, S. (2016). Comparing national environmental and economic performances through emergy sustainability indicators: Moving environmental ethics beyond anthropocentrism toward ecocentrism. *Renewable and Sustainable Energy Reviews, 58*, 1532–1542. https://doi.org/10.1016/j.rser.2015.12.188.

Macedo, F. M. F., Boava, D. L. T., Cappelle, M. C. A., & Oliveira, M. de L. S. (2012). Relações de gênero e subjetividade na mineração: Um estudo a partir da fenomenologia social TT - gender relations and subjectivity: A study from social phenomenolog. *Revista de Administração Contemporânea, 16*(2), 217–236. https://doi.org/10.1590/S1415-65552012000200004.

MacKinnon, K., Dudley, N., & Sandwith, T. (2011). Natural solutions: Protected areas helping people to cope with climate change. *Oryx*. https://doi.org/10.1017/S0030605311001608.

MacNaughton, G. (2017). Vertical inequalities: Are the SDGs and human rights up to the challenges? *International Journal of Human Rights, 21*(8), 1–23. https://doi.org/10.1080/13642987.2017.1348697.

Madurwar, M. V., Ralegaonkar, R. V., & Mandavgane, S. A. (2013). Application of agro-waste for sustainable construction materials: A review. *Construction and Building Materials, 38*, 872–878. https://doi.org/10.1016/j.conbuildmat.2012.09.011.

Masi, F., Rizzo, A., & Regelsberger, M. (2018). The role of constructed wetlands in a new circular economy, resource oriented, and ecosystem services paradigm. *Journal of Environmental Management, 216*, 275–284. https://doi.org/10.1016/j.jenvman.2017.11.086.

Maupin, A. (2017). The SDG13 to combat climate change: An opportunity for Africa to become a trailblazer? *African Geographical Review, 36*(2), 131–145. https://doi.org/10.1080/19376812.2016.1171156.

McClintock, B., & Bell, P. (2013). Australia's mining interests within Nigeria and Libya: Policies, corruption and conflict. *International Journal of Law, Crime and Justice, 41*(3), 247–259. https://doi.org/10.1016/j.ijlcj.2013.06.004.

McInnes, R. J. (2018). Sustainable development goals. In C Nick, et al. (Eds.), *The wetland book: I: Structure and function, management, and methods*. Springer.

McKay, V. (2018). Literacy, lifelong learning and sustainable development. *Australian Journal of Adult Learning*.

Melorose, J., Perroy, R., & Careas, S. (2015). Sustainable development goals FACT SHEET. *Statewide Agricultural Land Use Baseline, 2015*. https://doi.org/10.1017/CBO9781107415324.004.

Mendler de Suarez, J., Cicin-Sain, B., Wowk, K., Payet, R., & Hoegh-Guldberg, O. (2014). Ensuring survival: Oceans, climate and security. *Ocean and Coastal Management, 90*, 27–37. https://doi.org/10.1016/j.ocecoaman.2013.08.007.

Mkumbo, W. C. (2016). The role of School libraries in realizing the achievement of inclusive and equitable quality education in Tanzania: SDGs by 2030 an overview of sustainable development goals (SDGs). *International Researchers: Journal of Library and Information Science*.

Moh, Y. C., & Abd Manaf, L. (2014). Overview of household solid waste recycling policy status and challenges in Malaysia. *Resources, Conservation and Recycling, 82*, 50–61. https://doi.org/10.1016/j.resconrec.2013.11.004.

Mohieldin, M., & Caballero, P. (2015). Protect, restore and promote sustainable use of terrestrial ecosystems, sustainably manage forests, combat desertification, and halt and reverse land degradation and halt biodiversity loss. *UN Chronicle*. https://doi.org/10.18356/f405cab8-en.

Moralejo-Gárate, H., Kleerebezem, R., Mosquera-Corral, A., Campos, J. L., Palmeiro-Sánchez, T., & van Loosdrecht, M. C. M. (2014). Substrate versatility of polyhydroxyalkanoate producing glycerol grown bacterial enrichment culture. *Water Research, 66*, 190–198. https://doi.org/10.1016/j.watres.2014.07.044.

Müller, F., Jaeger, D., & Hanewinkel, M. (2019). Digitization in wood supply – a review on how industry 4.0 will change the forest value chain. *Computers and Electronics in Agriculture, 162*, 206–218. https://doi.org/10.1016/j.compag.2019.04.002.

Müller, J. M., Kiel, D., & Voigt, K. I. (2018). What drives the implementation of industry 4.0? The role of opportunities and challenges in the context of sustainability. *Sustainability, 10*(1). https://doi.org/10.3390/su10010247.

Nhamo, G., Nhemachena, C., & Nhamo, S. (2019). Is 2030 too soon for Africa to achieve the water and sanitation sustainable development goal? *Science of the Total Environment, 669*, 129–139. https://doi.org/10.1016/j.scitotenv.2019.03.109.

Oestreich, J. E. (2018). *SDG 10: Reduce inequality in and among countries*. Social Alternatives.

Onu, P., & Mbohwa, C. (2018a). Future energy systems and sustainable emission control: Africa in perspective. In *Proceedings of the International Conference on Industrial Engineering and Operations Management* (pp. 793–800). IEOM.

Onu, P., & Mbohwa, C. (2018b). Green supply chain management and sustainable industrial practices: Bridging the gap. In *Proceedings of the International Conference on Industrial Engineering and Operations Management* (pp. 786–792). IEOM (Washington DC).

Onu, P., & Mbohwa, C. (2018c). Sustainable oil exploitation versus renewable energy initiatives: A review of the case of Uganda. In *Proceedings of the International Conference on Industrial Engineering and Operations Management* (pp. 1008–1015). IEOM (Washington DC).

Onu, P., & Mbohwa, C. (2019a). Sustainable production: New thinking for SMEs. *Journal of Physics: Conference Series, 1378*(2). https://doi.org/10.1088/1742-6596/1378/2/022072.

Onu, P., & Mbohwa, C. (2019b). Sustainable supply chain management: Impact of practice on manufacturing and industry development. *Journal of Physics: Conference Series, 1378*(2). https://doi.org/10.1088/1742-6596/1378/2/022073.

Onu, P, & Mbohwa, C. (2019c). *Advances in solar photovoltaic grid parity*. IEEE.

Onu, P., & Mbohwa, C. (2019d). *Concentrated solar power technology and thermal energy storage : A brief overview of nascent sustainable designs*. IEEE.

Onu, P., & Mbohwa, C. (2019e). Renewable energy technologies in brief. *International Journal of Scientific and Technology Research, 8*(10), 1283–1289.

Onu., P., & Mbohwa, C. (2019f). Industrial energy conservation initiative and prospect for sustainable manufacturing. *Procedia Manufacturing, 35*, 546–551. https://doi.org/10.1016/j.promfg.2019.05.077.

Onu, P., & Mbohwa, C. (2020). An investigation into industrial manufacturing and sustainable production implementation in sub-Saharan Africa. *International Journal of Scientific and Technology Research*. In press.

Onyekwere, L. A. (2016). Enriching the role/importance of social workers in promoting peaceful and inclusive society for sustainable development. *International Journal of Scientific Research in Education, 9*(4), 212–220.

Owens, T. L. (2017). Higher education in the sustainable development goals framework. *European Journal of Education, 52*(4). https://doi.org/10.1111/ejed.12237.

Parisotto, A. (2015). Promote sustained, inclusive and sustainable economic growth, full and productive employment and decent work for all. *UN Chronicle*. https://doi.org/10.18356/9a54dfe4-en.

Pickin, J., & Randell, P. (2017). *Australian national waste report 2016*. Australian Government: Department of the Environment and Energy.

Pogge, T., & Sengupta, M. (2015). The sustainable development goals: A plan for building a better world? *Journal of Global Ethics, 11*(1), 1–9. https://doi.org/10.1080/17449626.2015.1010656.

Pogge, T., & Sengupta, M. (2016). Assessing the sustainable development goals from a human rights perspective. *Journal of International and Comparative Social Policy.* https://doi.org/10.1080/21699763.2016.1198268.

PWC. (2013). *Mining for talent: A study of women on boards in the mining industry by WIM(UK) and PWC.* PWC.

Rai, S. M., Brown, B. D., & Ruwanpura, K. N. (2019). SDG 8: Decent work and economic growth – a gendered analysis. *World Development, 113*, 368–380. https://doi.org/10.1016/j.worlddev.2018.09.006.

Recuero Virto, L. (2018). A preliminary assessment of the indicators for sustainable development goal (SDG) 14 "Conserve and sustainably use the oceans, seas and marine resources for sustainable development." *Marine Policy, 89*, 47–57. https://doi.org/10.1016/j.marpol.2018.08.036.

Revi, A., & Rosenzweig, C. (2013). The urban opportunity : Enabling transformative and sustainable development. *Agenda.*

Rice, J. L., Burke, B. J., & Heynen, N. (2015). Knowing climate change, embodying climate praxis: Experiential knowledge in Southern Appalachia. *Annals of the Association of American Geographers, 105*(2), 253–262. https://doi.org/10.1080/00045608.2014.985628.

Ringler, C., Willenbockel, D., Perez, N., Rosegrant, M., Zhu, T., & Matthews, N. (2016). Global linkages among energy, food and water: An economic assessment. *Journal of Environmental and Social Sciences, 6*, 161–171. https://doi.org/10.1007/s13412-016-0386-5.

Schroeder, P., Anggraeni, K., & Weber, U. (2019). The relevance of circular economy practices to the sustainable development goals. *Journal of Industrial Ecology.* https://doi.org/10.1111/jiec.12732.

Seth, N. (2015). Strengthen the means of implementation and revitalize the global partnership for sustainable development. *UN Chronicle.* https://doi.org/10.18356/3235a110-en.

Sheahan, M., & Barrett, C. B. (2017). Ten striking facts about agricultural input use in sub-Saharan Africa. *Food Policy, 67*, 12–25. https://doi.org/10.1016/j.foodpol.2016.09.010.

Siew, R. Y. J. (2015). A review of corporate sustainability reporting tools (SRTs). *Journal of Environmental Management, 164*, 180–195. https://doi.org/10.1016/j.jenvman.2015.09.010.

Singh Sekhon, B. (2014). Nanotechnology in agri-food production: An overview. *Nanotechnology, Science and Applications, 7*(2), 31–53. https://doi.org/10.2147/NSA.S39406.

Soliva, M., Bernat, C., Gil, E., Martínez, X., Pujol, M., Sabaté, J., & Valero, J. (2007). Education and research related to organic waste management at agricultural engineering schools. *International Journal of Sustainability in Higher Education, 8*(2), 1–10. https://doi.org/10.1108/14676370710726670.

Sousa, V., Almeida, N. M., & Dias, L. A. (2014). Risk-based management of occupational safety and health in the construction industry - part 1: Background knowledge. *Safety Science, 66*, 75–86. https://doi.org/10.1016/j.ssci.2014.02.008.

Tanaka, K. (2011). Review of policies and measures for energy efficiency in industry sector. *Energy Policy, 39*(10), 6532–6550. https://doi.org/10.1016/j.enpol.2011.07.058.

Tang, Q. (2015). Ensure inclusive and equitable quality education and promote lifelong learning opportunities for all. *UN Chronicle.* https://doi.org/10.18356/b2b87b7d-en.

Tchiehe, D. N., & Gauthier, F. (2017). Classification of risk acceptability and risk tolerability factors in occupational health and safety. *Safety Science, 92*, 138–147. https://doi.org/10.1016/j.ssci.2016.10.003.

Tewdwr-Jones, M. (2015). *Transforming our world: The 2030 agenda for sustainable development: Resolution adopted by the general assembly.* DISP.

UNESCO. (2016). *Incheon declaration and framework for action for the implementation of sustainable development goal 4: Ensure inclusive and equitable quality education and promote lifelong learning opportunities for all* (Report). United Nations Educational, Scientific and Cultural Organization.

UNIDO. (2017). *Structural change for inclusive and sustainable industrial development.* Vienna: United Nations Industrial Development Organization.

United Nations. (2014). *United nations decade of sustainable energy for all.* Retrieved July 5, 2018 from: http://www.un.org/millenniumgoals/pdf/SEFA.pdf.

United Nations. (2015). *Sustainable development knowledge plataform.* United Nations ESCAP.

United Nations. (2016a). *Goal 9: Build resilient infrastructure, promote sustainable industrialization and foster innovation.* Sustainable Development Goals.

United Nations. (2016b). *Sustainable Development GOALS - 17 goals to transform our world.* Sustainable Development Goals - United Nations. https://doi.org/United Nations Development Program (UNDP).

United Nations. (2018). *SDG goals 11: Make cities inclusive, safe, resilient and sustainable.* United Nations-DESA.

United Nations Development Programme. (2016). *Goal 10: Reduced inequalities.* Sustainable Development Goals.

United Nations Educational Scientific and Cultural Organization. (2013). *Education transforms lives.* United Nations Educational.

Vergragt, P., Akenji, L., & Dewick, P. (2014). Sustainable production, consumption, and livelihoods: Global and regional research perspectives. *Journal of Cleaner Production, 63*, 1–12. https://doi.org/10.1016/j.jclepro.2013.09.028.

Vuitton, D. A., & Dalphin, J. C. (2017). From farming to engineering: The microbiota and allergic diseases. *Engineering.* https://doi.org/10.1016/J.ENG.2017.01.019.

Waldron, K. (2009). Handbook of waste management and co-product recovery in food processing. In K. Waldron (Ed.), *Handbook of waste management and Co-product recovery in food processing.* Woodhead.

Wang, S. (2013). Reducing poverty through agricultural development in China. *IDS Bulletin.* https://doi.org/10.1111/1759-5436.12057.

WEC. (2017). *Energy trilemma index.* Retrieved July 6, 2018 from Monitoring the Sustainability of National Energy Systems website: https://www.worldenergy.org/wp-content/uploads/2017/11/Energy-Trilemma-Index-2017-Report.pdf.

Williams, T. O., Mul, M., Cofie, O., Kinyangi, J., Zougmore, R., Wamukoya, G., … Campbell, B. (2015). *Climate smart agriculture in the African context.* Cgspace.Cgiar.Org.

Xu, Y., & Szmerekovsky, J. (2017). System dynamic modeling of energy savings in the US food industry. *Journal of Cleaner Production, 165*, 13–26. https://doi.org/10.1016/j.jclepro.2017.07.093.

Yogo, T. U. (2017). Assessing the effectiveness of foreign aid in the education sector in Africa: The case of primary education. *African Development Review, 29*(3), 389–402. https://doi.org/10.1111/1467-8268.12276.

Yusuf, A. A., Onu, P., Hassanb, A. S., Tunji, A. L., Oyagbola, I. A., Mustafad, M. M., & Danjuma, A. Yusuf (2019). Municipality solid waste management system for Mukono District. *Procedia Manufacturing, 35*, 613–622. https://doi.org/10.1016/j.promfg.2019.06.003.

Zaman, A. U. (2015). A comprehensive review of the development of zero waste management: Lessons learned and guidelines. *Journal of Cleaner Production, 91*, 12–25. https://doi.org/10.1016/j.jclepro.2014.12.013.

Zimm, C. (2019). Methodological issues in measuring international inequality in technology ownership and infrastructure service use. *Development Studies Research*. https://doi.org/10.1080/21665095.2019.1605533.

CHAPTER

6 New approach and prospects of agrowaste resources conversion for energy systems performance and development

6.1 Introduction

There is an urgent need to have a reliable and efficient clean energy production processes and improvement in sustainability disposition beckons for better understanding (Onu & Mbohwa, 2019d). Recently, the exploitation of energy resources and the approaches used have become of quintessential importance. Hence, the need to determine ways in which they affect the environment, health, and safety of employees at any production/process stations cannot be overemphasized. These have led to the development of new energy technologies and waste to wealth initiatives that least affect social and environmental perspectives (Shedrack et al., 2019). While some of these energy technologies are generally thought to be safe for production of none carcinogen-based emissions (Akiyode, Tumushabe, Hadijjah, & Onu, 2017; Onu & Mbohwa, 2018), the operational activities and systems involved to execute the processes may either directly or indirectly impact climate change. Some of the essential operations include engineering materials selection, manufacturing equipment, and types of machinery, building facilities and energy options, transportation means, and energy sources, to mention just a few.

According to the International Energy Agency (2018) (IEA): the world consumed 2.1% as against 0.9% of the previous year, more energy than in 2016, which amounts to about 14,050,000,000 tons of oil equivalent (10,035,000 tons in 2000). Most recently, however, Asia shows unparalleled potential in energy consumption, with India and China leading the course, and contributing over 40% (combined) of the value initially reported. It has become clear that the development of renewables and its dedicated technologies in the areas of natural

Agricultural Waste Diversity and Sustainability Issues. https://doi.org/10.1016/B978-0-323-85402-3.00007-3

gas production contributes to the growth potential of modern electrification in rural and isolated communities (Kuang et al., 2016; Onu & Mbohwa, 2019a; Sgroi, Donia, & Alesi, 2018). This chapter focuses on biodegradable feedstock conversion for clean energy production; much emphasis is given to the process, type of fuels, characteristics, application, and significance of the energy resources potential of agricultural wastes.

Agro-wastes resources conversion and methane production potential

The euphoria for sustainable development is meant to promote social and environmental responsibilities, among all well-meaningful stakeholders. The focus, therefore, is on resource management and value creation from everyday operations to meet materials and energy demands (Abdul Qayoom Tunji Lawal et al., 2019; Abdulfatah et al., 2019; Onu, Abolarin, & Anafi, 2017). To this end, scientists concentrate their efforts on the selection of the best energy resources that are clean, readily available, economical, without the emission of noxious gases, or having any radioactive implication (Onu & Mbohwa, 2018). Biogas is one of the renewable nonfossil fuels that can be exploited from the wastes from agricultural operations to meet alternative bioenergy solutions. Biogas is obtained from the fermentation process of biomass containing carbohydrates with the help of microorganisms. The main content of biogas that serves as a fuel (biofuel) is methane gas (CH_4), also called biomethane. Table 6.1 demonstrates the biofuel production capacity from different feedstocks.

The nature of progress expected from any "agroresources conversion to energy project" depends on the methodologies, technologies, and the active participation of the concerned parties, including relevant government agencies and selection of the waste resources. Researchers have explored global food wastes along the supply chain (FLW Protocol, 2016; Papargyropoulou, Lozano, Steinberger, Wright, & Ujang, 2014). The result reveals that up to 30%, agricultural farm produce is lost during harvest, transportation, processing, and packaging. Hence, measures to recover this lot can contribute to the value chain of bioenergy production. More so, the operations involved in the waste collection process and the nature of pretreatment carried out on the wastes contribute to the characteristics of the biofuel that will be produced. Other factors such as higher production costs and lack of competitiveness with regard to the level of investment seen in fossil fuel industries can also affect the productivity and sustenance of the biodiversity industry (Onu & Mbohwa, 2019a,b). This chapter provides insights into the different biofuels and production techniques: anaerobic digestion, incineration, and gasification process, including the characteristic property of biofuels.

Table 6.1 Different agricultural wastes/feedstock and their biofuel potential.

Types of resources	Agrowaste	Bio fuels				References
		Bioethanol yield	Biomethane yield	Biobutanol yield	Biohydrogen yield	
Crop residue	Rice straw	12–29.1[a]	302[c]	Biobutanol yield	24.8[g]	Akhtar, Goyal, & Goyal, 2017; Asadi & Zilouei, 2017; Chandra, Takeuchi, & Hasegawa, 2012; Gottumukkala et al. (2013)
	Wheat straw	11.8–71.2[a]	290[c]	Biobutanol yield	1–68[f]	Chandra, Takeuchi, & Hasegawa, 2012; Guo, Trably, Latrille, Carrre, & Steyer, 2010; Qureshi, Cotta, & Saha, 2014; Talebnia, Karakashev, & Angelidaki, 2010
	Corn stover	40–55.27[a]	338[c]	Biobutanol yield	49–66[f]	Guo, Trably, Latrille, Carrre, & Steyer, 2010; Lau & Dale, 2009; Liew, Shi, and Li (2012); Qureshi, Cotta, & Saha, 2014
	Barley straw	11.9–46.0[a]	–	Biobutanol yield	–	García-Aparicio et al. (2011); Han, Kang, Kim, & Choi, 2013; Saha & Cotta, 2010; Qureshi, Cotta, & Saha, 2014
	Sugarcane bagasse	23.2–59.1[a]	278[c]	Biobutanol yield	6980[h]	Cardona, Quintero, & Paz, 2010; Hu, Li, Wang, & Zhu, 2018; Pang et al. (2016); Sangyoka, Reungsang, & Lin, 2016
Agroindustrial waste resources	*Pongamia pinnata* deoiled seed cake	0.088[b]	448[c]	–	–	Chandra, Takeuchi, & Hasegawa, 2012; Doshi & Srivastava, 2013
	Rice bran	–	–	2.75[c]	545[h]	Chandra, Takeuchi, & Hasegawa, 2012; Taslim, Bani, Parinduri, & Ningsih, 2018; Tandon, Thakur, Tiwari, & Jadhav, 2018; Zhang, Wong, & Yung, 2013

Continued

Table 6.1 Different agricultural wastes/feedstock and their biofuel potential.—*cont'd*

Types of resources	Agrowaste	Bio fuels				References
		Bioethanol yield	Biomethane yield	Biobutanol yield	Biohydrogen yield	
	Apple pomace	8.44[b]	–	12.92[e]	134.04[g]	Gupta & Verma, 2015; Hijosa-Valsero, Paniagua-García, & Díez-Antolínez, 2017; Wang, Wang, Fang, Wang, & Bu, 2010
	Orange peel	6[b]	217[c]	19.5[d]	–	Boluda-Aguilar & López-Gómez, 2013; Joshi, Waghmare, Sonawane, & Waghmare, 2015; Wikandari, Nguyen, Millati, Niklasson, & Taherzadeh, 2015
	Rice bran deoiled cake	–	–	5.16[d]	295[h]	Al-Shorgani, Kalil, & Yusoff, 2012; Tandon, Thakur, Tiwari, & Jadhav, 2018
	Oil palm fruit bunch	37.8[h]	276–340	1.262–2.61[d]		Ibrahim, Abd-Aziz, Razak, Phang, & Hassan, 2012; Kim & Kim, 2013; Noomtim & Cheirsilp, 2011; O-Thong, Boe, and Angelidaki (2012)
Livestock resources	Cattle manure	–		–	65 [f]	Guo, Trably, Latrille, Carrre, & Steyer, 2010; Li, Dang, Zhang, Zou, and Yuan (2015)
	Pig manure	–		–	14–18[f]	Guo, Trably, Latrille, Carrre, & Steyer, 2010; Li, Liu, & Sun, 2016

[a] *L/kg of dry biomass.*
[b] *%.*
[c] *L/kg VS.*
[d] *g/L ABE.*
[e] *g/L.*
[f] *mL H_2/g VS.*
[g] *mL H_2/g TS.*
[h] *mL H_2/L.*

6.2 Biofuels production from agricultural wastes

Production of biofuels from waste agricultural biomass

Bioethanol

The potential of biofuel development is beneficial to support the transport industry, which is apt to rethink strategies that contribute to climate change mitigation and, hence, to embrace ethanol production of clean and alternative energy from available biodegradable or agricultural wastes. The development of several techniques and advancements in ethanol production has seen a geometric growth both globally and in Africa, since the past decade. The worldwide acceptance of bioethanol fuel and its output is due to the properties of the fuels, having high octane and oxygen content that allows it to ignite and burn quickly and compares favorably with petrol. Regardless of the minute amount of NOx and particulate matter released during the burning of bioethanol, the fuel reduces Cox emission by more than 79% when compared with other conventional automobile fuels. Research has shown that bioethanol is found useful and can be blended with petrol, such as biofuel production from agricultural biomass (in a million tons): Corn stover (731.3 MT), rice (354.34 MT), and wheat straw (128.02 MT) yielded (in gigaliters) 205, 104, and 58.6 GL of bioethanol, respectively (Saini, Saini, & Tewari, 2015).

The tendency for higher bioethanol production is dependent on the cellulose and hemicellulose composition. In essence, the presence of lignin hinders the hydrolysis and conversion (biological) process of ethanol production. More so, the characteristics of most agrowastes/residues, due to ash and silica, of the alkali and protein contents as in rice and wheat straws make them less viable to undergo the thermochemical process, thereby resulting in lesser biofuel (ethanol production) and thus resulting in the compulsory process where the biomass feedstock is pretreated to improve the yield of bioethanol. While alkaline pretreatment has been known to be an effective strategy for the separation of lignin and aid in the removal of silica, the process is sufficient to control the formation and facilitate the removal of complex and unwanted compounds, which affects the high production yield of ethanol (Yuan et al., 2018).

Qiu et al. (2017) have experimented on the propensity of ethanol production from the combination of different chemicals over the use of alkali alone, for wheat straw. Furthermore, ethanol production from corn stover under the combined pretreatment, involving alkaline and ammonia, yielded far greater ethanol content compared with the treatment with only fungi, which was investigated by Zhao, Damgaard, & Christensen, 2018. One of the most preferable and widely studied yeast strains for ethanol production through fermentation is *S. cerevisiae*, with thermophilic nature that can cause a high yield of ethanol. The production of bioethanol from sugarcane (180.73 MT) has also been investigated, yielding 51.3 GL, according to Saini, Saini, & Tewari, 2015. Although sugarcane bagasse biomass has drawn interest and is seemed to be one of the most readily available and viably preferable crop residues as feedstock for bioethanol production, its yield in terms of bioethanol production is considerably low, despite the essential pretreatment processes being followed.

Bioethanol production from the food and processing industry now explores energy crops and substrates of some deoiled seed (roughages) and fruit peels (Abu Tayeh, Najami, Dosoretz, Tafesh, & Azaizeh, 2014; Cardona, Quintero, & Paz, 2010; Mendonça et al., 2019). It is observed that the categories of roughages/cakes, which contain proteins and are high fiber content, are invariably potent with cellulose, hemicellulose, and lignin compounds and form the building block of resources for bioethanol production. Moreover, several noncereal biomass and crop residues that are considered for sustainable bioethanol production are expected to overcome the challenges of a limited supply of the feedstock, technicality to handling the wastes, and pretreatment criteria to ensure absolute degradation of lignin and optimal biofuel production (Gupta & Verma, 2015).

Biogas

Biomethane (biogas) is one of the commonest attractions for the conversion and uses of agrowastes for clean energy proposition and the production of biofuels. The presence of cellulose, hemicellulose, lipids, protein, and carbohydrates gives agricultural residues their potential to be explored as an energy crop and for the production of green fuels. The higher the lipid contents, the greater the tendency for methane to be produced (higher energy content), compared with carbohydrate and protein constituent, being potential energy feedstock. Although the production of methane has been deemed feasible and successfully carried out over the years, an investigation into animal waste proof to be a more suitable feedstock for biogas production. This is due to the high moisture and organic content and low carbon to nitrogen ratio (C:N), which causes the feedstock (animal dung) to be easily fermented. Moreover, research has indicated that the substrates to inoculum ratio (S:I) of animal feedstock are suitable under normal temperature condition (psychrophilic (20°C), mesophilic (32 to 37°C), and thermophilic (55°C)) as a monosubstrate for the generation of the satisfactory yield of biogas (Ren et al., 2018; Shahbaz, Ammar, Zou, Korai, & Li, 2019).

However, biogas production through cogeneration, involving animal wastes (dung) and agrowaste biomasses, has shown excellent yield in the production of biofuels, compared with other high cellulose fruit peels and cereal plants. The presence of nitrogen in animal waste is an essential factor, so as the sugar content in the plant/fruit residues, thus promoting bacteria activities that facilitate the anaerobic process. Regardless, in a bid to further improve the biodegradability and improve the yield of biofuel production, a combination of nitrogenous-based wastes with plant/fruit wastes has been successfully demonstrated (Scano et al., 2014).

The different challenges that affect anaerobic digestion process includes the formation of unwanted radicals (ammonia and sulfur) and organic compounds that causes failure of the operation (Shahbaz, Ammar, Zou, Korai, & Li, 2019; Shi et al., 2017). As such, ammonia inhibitors are controlled through the adjustment of the C:N and the acidity (pH) of the biofuel production arrangement (Wang, Lu, Li, & Yang, 2014). Also, an extraordinary approach to handle the ammonia inhibition is through increasing the rate of organic loading or by an airstrip technique, using inert metals and clay (Paul & Dutta, 2018; Zeshan, Karthikeyan, & Visvanathan, 2012). Sulfide inhibition, on the other hand, can be controlled when an optimal ratio of carbon and sulfide (C:S) is maintained. More so, the pretreatment process, which yields HMF and furfural, offers an alternative and more preferred path to resolving the challenges of ammonia and sulfide and, hence, the proposal to use acid and thermal approaches, i.e., hydrothermal pretreatment process (Paul & Dutta, 2018).

Biobutanol

The significance of butanol as a clean energy fuel ranges from its energy content, which is considered to be higher than ethanol, with relatively low volatility, flammability, and vapor pressure characteristics, and even beats the desire to be used, in place of biodiesel (Geng, Cao, Tan, & Wei, 2017; Li, Tang, Chen, Liu, & Lee, 2019; Szulczyk, 2010). The clean biofuel has been found to have a wide variety of applications: used as an additive and blended with other synthetic and natural fuels to improve the performance of the system. The diversification of agroindustrial and farm wastes as biomass resource for the production of biobutanol opens a new chapter in sustainability and advancement of biorefinery technology (Pereira, Dias, Mariano, Maciel Filho, & Bonomi, 2015). The acetone–butanol–ethanol (ABE) fermentation process has been known to be one of the most optimal processes/paths to produce a high yield of biobutanol using high carbohydrate (starch and glucose) substrates (Zhou et al., 2014).

In an experimental case, the use of sugarcane bagasse, in the fed-batch fermentation process (*Clostridium acetobutylicum* GX01), Pang et al. (2016) reported the production of ABE in the range of 20%–30%. Other investigations considered the *C. acetobutylicum* XY16 fermentation process, which yielded acetone (29%), butanol (65%), and ethanol (6%) (Kong et al., 2016). Furthermore, wastes/chaffs from barleys and corn stover as agricultural waste have been explored for production of butanol through the use of *Clostridium beijerinckii* P260, showing significance and yield above 45 g/L of ABE from the feedstocks (Qureshi, Cotta, & Saha, 2014). From general knowledge, the choice of pretreatment operation and nature of enzymes/catalytic process, which acts on the feedstock, will impact the yield of biobutanol, compared with other nontreated biomass wastes (Al-Shorgani, Kalil, & Yusoff, 2012; Smith et al., 2015). In recent times, the exploitation of an ordinary day-to-day food crop biomass (lettuce) has been investigated for biobutanol potential. Although, due to the high carbohydrate content in lettuce, the yield of the fuels has been found to be low (Procentese, Raganati, Olivieri, Elena Russo, & Marzocchella, 2017), it confirms the potential to generate gainful energy from wastes coming from farming operations (discards during harvest) or preparation of the farm produce.

Biohydrogen

Biohydrogen is a clean energy fuel produced from the conversion of easily biodegradable agroresources. It is a form of green energy and, with each molecule of the hydrogen having high enough energy that allows it to be used as a fuel, thereby, produces fluid (water) after combustion (Di Paola, Russo, & Piemonte, 2015). The discovery and development of hydrogen technology have brought new hope for renewable energy expansion to regularize global warming effects, caused by pollution from burning fossil fuels (Dalena, Basile, & Rossi, 2017). The processes involved in the production of biohydrogen require organic biodegradable compounds, used as substrate either during the autotrophic conversion process, where photosynthetic consideration plays a pivotal role. In such a case, the use of certain organisms (algae, protists, and other single-celled microbes) that can facilitate the conversion of the sun's energy to hydrogen directly is required. Also, the heterotrophic conversion process, involving the breakdown of carbohydrate contents of biomass feedstocks (agrowastes), in the presence of anaerobes through a dark fermentative conversion process, could be another option (Ghimire et al., 2015).

Understandably, the biohydrogen production process follows a complex series of activities and is limited by the heterogeneous nature of the substrate being used, whether of complex structure and preventing microbial activities (hydrolysis) or not. This is backed by previous research, on the pretreatment of corn stalk for the production of biohydrogen, which seems to increase yield in both acid and alkaline treatment, compared with other pretreated feedstocks used (Bala-Amutha & Murugesan, 2013; Nasr et al., 2014). Further research shows that the nature of pretreatment used to process biomass feedstocks affects the quality of the hydrogen that will be generated. Yin et al. (2014) investigated the production of biohydrogen from two different groups of agrowastes: livestock/manure and crop residues. In their resolve, they conclude with the assertion that the production of ammonia in the biorefining process leads to reduced productivity of hydrogen from livestock wastes when compared with the case of the crop biomasses. Moreover, when the environmental condition becomes unfavorable for microbial activities, it hinders biohydrogen generation.

It is worth to note that under normal conditions, the operational temperature requirement for anaerobic operation (thermophilic condition), coupled with necessary physicochemical pretreatment conditions, can impact biohydrogen yield from animal waste/manure, whereas in a cogeneration process, combining the livestock (rich in nitrates and carbohydrates) and crop residues waste will result in higher production of biohydrogen (Wu, Yao, & Zhu, 2010). Hence, the wide varieties of agrowastes, second generation of crops high in carbohydrates/simple sugar, and agroindustrial food processed wastes (whey of brewing or cheese production waste) are potential resources for the generation of gainful biohydrogen yields (Guo, Trably, Latrille, Carrre, & Steyer, 2010; Urbaniec & Bakker, 2015).

Biodiesel

The chances of extracting sufficient amount of biodiesel from agrowastes, and farm produce, including vegetables, cereal wastes, rice bran, and so on, or wastes from agroindustrial food processing industry operation (roughages of oil seeds, food processing/wastewater, and food discarded or unwanted foods) have been researched over the years (Gupta & Demirbas, 2010; Stephen & Periyasamy, 2018). Other wasteful materials such as cooking oils, as well as dietary and animal wastes, have been identified as potential feedstocks for the production of clean, renewable biodiesel fuel (Karmakar & Halder, 2019). The use of a sustainable environmentally friendly solvent has been demonstrated as suitable for the extraction of biooil from waste generated from coffee, which is a significant cash crop traded in sub-Saharan Africa, thus using dimethyl ether (DME), to yield about 16% of biodiesel (transesterification process) (Matzen & Demirel, 2016; Sakuragi, Li, Otaka, & Makino, 2016). Also, the global attention toward the utilization of rice bran for biodiesel production draws attention to its high oil content and the massive potential to be used for sustainable production of biofuels (Lin, Ying, Chaitep, & Vittayapadung, 2009; Patil, Kar, & Mohapatra, 2016).

The production of biodiesel from rice bran oil has been demonstrated through a transesterification process, which yielded an average of 60%−85% of biodiesel under favorably controlled conditions (Kattimani, Venkatesha, & Ananda, 2014). Other strategies considered for biodiesel production, using a catalyst to facilitate the process to go beyond 98% yield rate, have been investigated, and authenticated, using K_2CO_3-modified zeolite (Taslim, Bani, Parinduri, & Ningsih, 2018) and $HClSO_3-ZrO_2$-modified zirconia (Zhang, Wong, & Yung, 2013). Al-Hamamre (2015) researched about biodiesel production from roughages and wastes from oilseed crops (cakes), which yielded between 40% and 65% of the clean fuel, in a transesterification reaction process.

6.3 Physicochemical properties of biofuels

The different classes of organic wastes from which biofuels are extracted affect the characteristics and features of the fuel, which determines their areas of application and operational requirements. Hence, the need to ascertain the physical and chemical properties and to understand the criteria, which guide their usage, cannot be undermined.

Density

One of the most desirable properties of biofuels is their ability to provide an optimum power output, which is based on its flow propensity. In essence, it affects performance and emission (smoke formation) in an automotive engine. Also, when the density of biodiesels is compared closely with fossil diesel fuel, this property is affected by the origin of the feedstock. The following European, American, and international standards guide the definition, classification, and procedural measurement for biodiesel density: EN 14214, ASTM D6751, and ISO 12185:1996 (EN ISO 12185: 1996, 2019; Munari, Cavagnino, & Scientific, 2009).

Cold point

The solidification of biofuel at a certain specific temperature gives meaning to the term 'cold point.' It is the point when the fuel crystallizes into solid. It is not desirable for biofuels to clog in the automotive system. The increase in a measure of saturated fatty acid content in biofuels has a direct implication on the cloud point property, which, in turn, threatens the smooth operational performance of the engine.

Pour point

This is the minimum temperature before which a fuel transforms into a semisolid state. This property has a significant influence on the flow property of the fuel and is a measure of the fuel gelling point. The reference pour point temperature usually falls below the cloud point. The ASTM D6751 standard (ASTM D6751-15c, 2010) offers information about the classification and specification of different categories of methyl ester.

Boiling point

The property of a fuel, which tends to a low boiling point, makes it volatile and easy to evaporate. The grades of gasoline fuels that exist have low boiling points compared with diesel, which, on the contrary, has high boiling points. However, the potential of low boiling points in some categories of fuels and bioethanol promotes the instantaneous release of energy. This property is desired to maintain a balance between the challenges of gas leaks and incomplete combustion in engine cylinder heads.

Kinematic viscosity

To determine the quality of fuel, the kinematic viscosity test is conducted. This property puts into perspective the relationship between the frictional force and molecules of the fuel, and its tendency to undergo resistive flow. If the viscosity is high, the atomization capability of the fuel will diminish, leading to incomplete burning of the fuel and deposition of soots in the surrounding. Hence, it is the fluidity requirement of a specification of fuel, which must be high enough to support lubrication at considerable high-temperature ranges without disintegration and still withstand cold start conditions. The kinematic viscosity measurement, according to numerous (European (2013) and ASTM International (2015)) standards, refers to the 40°C (temperature) to determine the specification of fuels or ethyl ester that is measured. The hydrocarbon chain length is known to contribute to the viscosity value. As such, the long-chain in free fatty esters gives them a preferable viscosity property, compared with fossil fuels.

Calorific value (kJ/kg)

This is the amount of heat released during combustion when a unit of fuel is burnt. It affects the combustion characteristics, thermal efficiency, and rate consumption of the fuel. The measure of calorific value of biofuel can be determined through the DIN 51900-1 and is dependent on the fatty acid content and the presence of hydrogen, oxygen, and carbon water. The higher the calorific value, the less desirable the fuel, and with the majority of biofuels having lesser values due to higher oxygen content, alcohols are proved to be more desired compared with gasoline—fossil fuels (Zhang, Jacobson, Björkholtz, Munch, & Denbratt, 2016). However, the major pitfalls due to low calorific value are miscibility, unstable, and poor lubricity tendency, which can be improved through the use of additives (He, Chen, Zhu, & Zhang, 2018).

Saponification value

The saponification value is defined as "the number of milligrams of KOH needed to neutralize the fatty acids obtained by complete hydrolysis of 1 gram of an oil sample." It is essential to consider the relative increase in the average molecular weight of fuel, coming from a particular feedstock, which has a direct implication on the saponification value. However, these may be a result of the nature of the oil in the form of fatty acid and methyl ester or triglycerides.

Iodine value

When it is desired to ascertain the value of unsaturation in fatty acids, which represents the iodine value, or the unit mass of iodine (g) that is absorbed by 100 g of the chemical solvent, the methodology in preparation and testing must conform to the regulation in EN 14111:2003, as in the case of iodine value (IV). Thus, it represents the number of double bonds in a specified mass of oil (g), which reacts with a unit mass of iodine. It means to say that increasing the number of double bonds directly affects the rate of conversion and transesterification and the iodine value as well.

Cetane number

Cetane number expresses the ignition potential of fuel, as to which different engines have desired specifications that guide their selection of fuels in conformity with the design of the cylinder compartment. It is essential to test the combustion ability of fuel and ignition quality with perspective to the compression ratio. While the cetane number attributes the class to biodiesel fuel, an increase in its value will mean an increase in the carbon chain length of the fuel and is essential to improve operational stability and engine efficiency. As the cetane number increases, it leads to the slow burning of the fuel, by shortening of the ignition delays and maximizing fuel consumption. Furthermore, the higher cetane number facilitates the cold starts, with considerably less operating engine noise and exhaust gas emission. The implication is that there will need to redesign for ignition timing, which is dependent on the cylinder temperature and pressure and bases on the model of the engine, as modern engines now consider all the emission effects due to cetane number.

Acid value

The acid value is a measure of potassium hydroxide (mg) that is required to neutralize the free fatty acid contained in unit mass (g) of chemical substance. It is the property of a biofuel, which indicates the quality of lubricity or rate of degradation when stored over some time, with inference to stability and shelf life. The standardization of acid value of oil according to EN 14104:2003, with a benchmark value of 0.5 mg KOH/g, which also relates to ASTM D6751, is used to monitor the degradability of biodiesel. This value is higher, compared with fossil-based minerals, which implies a higher risk to cause corrosion defects.

Oxygen content

It is desired to achieve clean burning of fuel for efficient engine operation, thus requiring the use of alcohols rich in oxygen, so as to facilitate or speed up heating in the combustion chamber and the engine cylinder head. The oxygen level in biofuels is higher than that in diesel and gasoline, with the most oxygen content discovered in methanol. In essence, alcohol is preferred due to the propensity of higher combustion efficiency, without a direct impact on the fuel consumption ascribed to the presence of oxygen.

Water content

Under the production process of biofuels from organic feedstocks, it results in the presence of a high quantity of water, compared with mineral fuels. As such, it is detrimental to lead several setbacks and negatively affects the system's performance, leading to poor lubricity and high wear and corrosion tendencies. Also, the presence of water is disadvantageous to the performance of additives, leading to the reduced operational effectiveness of the oil or biodiesel. Internationally recognized standards that are used to guide water content measurements and limits are as follows: EN 14214 and EN ISO 12937.

Oxidative stability

This is a chemical process where oxygen is dissolved in oil, facilitated by catalysts, acidic medium, or elevated temperature. The characteristic features of oxidative stability have a significant impact on the viscosity and lubricative capability of a fatty acid methyl ester to improve engine performance. The degree of saturation of a chemical compound also has an impact on the oxidative stability; it increases as saturation is decreased. More so, the presence of an additive causes the biodiesel to be stable and overcome the limitation of storage over a long period. The European

EN 14214 standardization provides information concerning the determination of oxidative stability of fatty acid methyl ether.

Latent heat of vaporization

The latent heat of vaporization is the measure of the energy required when a substance is transformed from its semisolid/liquid state to gaseous vapor in the absence of any significant input or change in temperature. The rate of ignition of a fuel may be greatly influenced by this property and, thus, when the value is high, can lead to decrease in the operating temperature within the cylinders compartment (increase ignition delay and cold start tendencies) (Vallinayagam, Vedharaj, Yang, Roberts, & Dibble, 2015; Wang et al., 2016). Although mineral diesel has values of latent heat of vaporization, which are compared closely with butanol and other alcohols, the decision is made based on NOx and soot emission coupled with the most desirable option for engine operational efficiency (Gautam & Agarwal, 2015).

Autoignition temperature

This is the lowest temperature at which a flammable chemical substance under normal condition ignites spontaneously in the absence of a source of ignition. It is the occurrence of spark flames, characterized by different minerals and biofuels at which specific temperature the molecules reach an activation energy level and combustion ensues automatically. The autoignition temperatures of butanol, methanol, and ethanol are higher than diesel (230°C), which makes it less risky, with regard to storage or when transported.

6.4 Enhanced biofuel production: outlook and prospects

For centuries, humans had generated energy from various sources to satisfy the industry needs. These processes are associated with several environmental and climate change challenges, affecting habitation on land and sea, thus threatening the ecosystem. The decision to choose from the energy technologies that least affect the environment, since it has been proven that not all energy technologies contribute to greenhouse gas emission (global climate challenges) (Tie & Tan, 2013). However, the means to acquire the different engineering materials, the process of equipment production, machinery services, and building of facilities, among others, directly or indirectly affect the ecosystem (Onu & Mbohwa, 2019a,b). As such, the aforementioned actions unavoidably contribute to greenhouse gas emissions and global climate changes. Regardless, the gains outweigh the loss, as the long-term implication of sustainable and renewable energy development is vital to proffer solutions to these environmental challenges, as well as lead social and economic contributions.

The need for clean and efficient energy systems development and a sustainable approach to enhance the techniques for biofuels production is to ensure energy security. From the perspective of the utilization of wastes from farming and food production activities, different parameters can significantly affect the outcome of biofuel production (Abdul Kapor, Maniam, Rahim, & Yusoff, 2017). These include, but are not limited to, quality of feedstocks, free fatty acids, water content, temperature condition (effective reaction), time of response (availability), and type of catalysts required, among others. Several processes are involved, as well. The process of transesterification—acidic—involves the presence of sulfuric acid as a catalyst that impacts negatively on the production of biofuels, owing to slow reaction, which leads to poor conversion (De Lima, Ronconi, & Mota, 2016). Hence, That is, for optimal production of biofuels, the use of base-catalyzed transesterification process, will result in a high conversion rate (Colombo, Ender, & Barros, 2017).

Additionally, it is paramount to take note of the amount, in percentage proportion of the free fatty acid present (not to exceed 2.5%), to saponification that is not desired (Sharma & Singh, 2010). More so, Chen et al. (2017), has recommended that the anaerobic digestion process should not exceed 50% in operational performance and energy production. Hence, different protocols to overcome the drawback mentioned earlier extend to the conversion of the solid wastes (digestate) for energy purposes, through the pyrolytic and gasification processes, or following the liquefaction process (thermochemical conversion of biomass). The processes mentioned are extensively discussed in Chapter 7, highlighting their performances, limitations (ash production causing slag or bad smell), significance, and areas of application, based on different feedstock requirement. On the other hand, the conversion of liquid digestate finds a wide variety of usage through a direct liquefaction route for the production of viable, value solvents, described in Fig. 6.1 (conversion of wet biomass).

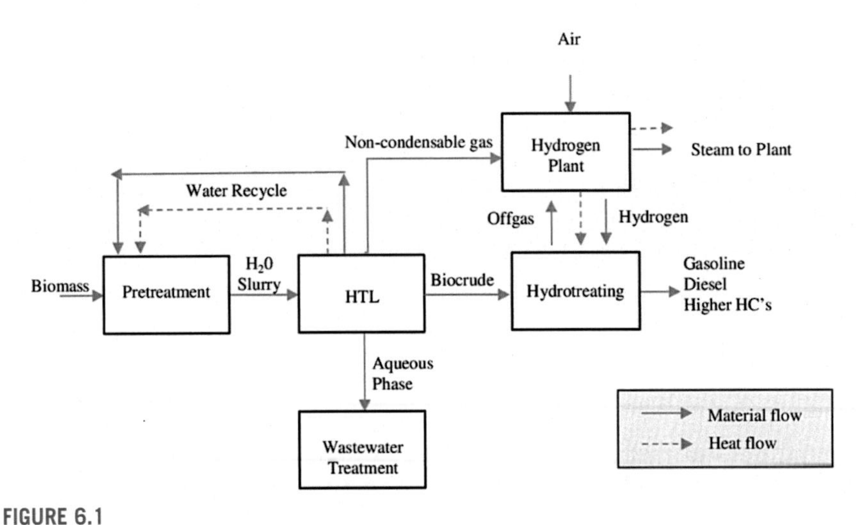

FIGURE 6.1

Hydrothermal liquefaction and hydrotreating process flow diagram.

6.5 Conclusion and perspectives

This chapter has described the different techniques and strategies of biofuel production with inference to the classification and characteristics of the fuels. Essential contributions to the understanding of the properties of biofuels have been presented. Trending research in sustainable and clean energy, while focusing on the improvement of physical and chemical processes, which are cost-effective for the production of biofuels is featured in this chapter. Although the majority of the processes are associated with various environmental issues linked to climate change, land, sea, and air pollution, and environmental safety initially deemed as a minor issue, they have become stepping stones to greater global challenges, whereas the best strategies to avoid the effects of agrowastes is the total avoidance of its production, and complete neutralization once they surface. The potential of the resources as feedstock, or as sources for clean energy, has been exploited to describe the types of fuels and characteristics of these fuels.

The production of bioethanol, biogas, biohydrogen, biobutanol, and biodiesel has received insight into this chapter. The perspectives for utilization of the various biofuels is underpinned to their capabilities and characteristic feature (physical and chemical), and properties which they exhibit, to determine their performance over conventional fossils (fuels and oils) that are in existence. It is essential to have biofuels that meet satisfactory quality standards and have specific properties that are consistent with available crude fuels or of better performance. Hence, the result of successfully producing biofuels from agrowaste resource must be followed by monitoring the fuel properties and possible techniques adapted to improve the operations process for commercial-scale production.

References

Abdul Kapor, N. Z., Maniam, G. P., Rahim, M. H. A., & Yusoff, M. M. (2017). Palm fatty acid distillate as a potential source for biodiesel production-a review. *Journal of Cleaner Production, 146*, 1–9. https://doi.org/10.1016/j.jclepro.2016.12.163.

Abdul, Qayoom Tunji Lawal, Ninsiima, E., Odebiyib, O. S., Hassan, A. S., Oyagbola, I. A., Onu, P., & Danjuma, A. (2019). Effect of unburnt rice husk on the properties of concrete. *Procedia Manufacturing, 35*, 635–640. https://doi.org/10.1016/j.promfg.2019.06.006.

Abdulfatah, Abdu Yusuf, Onu, P., Hassanb, A. S., Tunji, A. L., Oyagbola, I. A., Mustafad, M. M., & Danjuma, A. Yusuf (2019). Municipality solid waste management system for Mukono District. *Procedia Manufacturing, 35*, 613–622. https://doi.org/10.1016/j.promfg.2019.06.003.

Abu Tayeh, H., Najami, N., Dosoretz, C., Tafesh, A., & Azaizeh, H. (2014). Potential of bioethanol production from olive mill solid wastes. *Bioresource Technology, 125*, 24–30. https://doi.org/10.1016/j.biortech.2013.10.102.

Akhtar, N., Goyal, D., & Goyal, A. (2017). Characterization of microwave-alkali-acid pretreated rice straw for optimization of ethanol production via simultaneous saccharification and fermentation (SSF). *Energy Conversion and Management, 141*, 133–144. https://doi.org/10.1016/j.enconman.2016.06.081.

Akiyode, O., Tumushabe, A., Hadijjah, K., & Onu, P. (2017). Climate change, food security and environmental securi-ty: A conflict inclination assessment of karamoja region of Uganda. *International Journal of Scientific World.* Article 8458. https://doi.org/10.14419/ijsw.v5i2.8458.

Al-Hamamre, Z. (2015). Potential of utilizing olive cake oil for biodiesel manufacturing. *Energy Sources, Part A: Recovery, Utilization, and Environmental Effects, 37*(23), 2609–2615. https://doi.org/10.1080/15567036.2012.724758.

Al-Shorgani, N. K. N., Kalil, M. S., & Yusoff, W. M. W. (2012). Biobutanol production from rice bran and de-oiled rice bran by Clostridium saccharoperbutylacetonicum N1-4. *Bioprocess and Biosystems Engineering, 35*, 817–826. https://doi.org/10.1007/s00449-011-0664-2.

Asadi, N., & Zilouei, H. (2017). Optimization of organosolv pretreatment of rice straw for enhanced biohydrogen production using Enterobacter aerogenes. *Bioresource Technology, 227*, 335–344. https://doi.org/10.1016/j.biortech.2016.12.073.

ASTM D6751-15c. (2010). Standard specification for biodiesel fuel blend stock (B100) for middle distillate fuels. *ASTM International.* https://doi.org/10.1520/D6751.

Bala-Amutha, K., & Murugesan, A. G. (2013). Biohydrogen production using corn stalk employing Bacillus licheniformis MSU AGM 2 strain. *Renewable Energy, 50*, 621–627. https://doi.org/10.1016/j.renene.2012.07.033.

Boluda-Aguilar, M., & López-Gómez, A. (2013). Production of bioethanol by fermentation of lemon (*Citrus limon* L.) peel wastes pretreated with steam explosion. *Industrial Crops and Products, 41*, 188–197. https://doi.org/10.1016/j.indcrop.2012.04.031.

Cardona, C. A., Quintero, J. A., & Paz, I. C. (2010). Production of bioethanol from sugarcane bagasse: Status and perspectives. *Bioresource Technology, 101*, 4754–4766. https://doi.org/10.1016/j.biortech.2009.10.097.

Chandra, R., Takeuchi, H., & Hasegawa, T. (2012). Methane production from lignocellulosic agricultural crop wastes: A review in context to second generation of biofuel production. *Renewable and Sustainable Energy Reviews, 16*(3), 1462–1476. https://doi.org/10.1016/j.rser.2011.11.035.

Chen, G., Guo, X., Cheng, Z., Yan, B., Dan, Z., & Ma, W. (2017). Air gasification of biogas-derived digestate in a downdraft fixed bed gasifier. *Waste Management, 69*, 162–169. https://doi.org/10.1016/j.wasman.2017.08.001.

Colombo, K., Ender, L., & Barros, A. A. C. (2017). The study of biodiesel production using CaO as a heterogeneous catalytic reaction. *Egyptian Journal of Petroleum, 26*(2), 341–349. https://doi.org/10.1016/j.ejpe.2016.05.006.

Dalena, F., Basile, A., & Rossi, C. (2017). Bioenergy systems for the future: Prospects for biofuels and biohydrogen. In F. Dalena, et al. (Eds.), *Bioenergy systems for the future: Prospects for biofuels and biohydrogen*. Woodhead.

De Lima, A. L., Ronconi, C. M., & Mota, C. J. A. (2016). Heterogeneous basic catalysts for biodiesel production. *Catalysis Science and Technology*, (9). https://doi.org/10.1039/c5cy01989c.

Di Paola, L., Russo, V., & Piemonte, V. (2015). Membrane reactors for biohydrogen production and processing. In L. Di Paola, et al. (Eds.), *Membrane reactors for energy applications and basic chemical production*. Woodhead.

Doshi, P., & Srivastava, G. (2013). Sustainable approach to produce biocthanol from Karanja (Pongamia pinnata) oilseed residue. *Turkish Journal of Agriculture and Forestry, 37*(6), 781–788. https://doi.org/10.3906/tar-1207-18.

EN ISO 12185 : 1996. (2019). *EN ISO 12185 : 1996 | crude petroleum and petroleum PR... | SAI global*. Retrieved October 2, 2019, from https://infostore.saiglobal.com/en-gb/Standards/EN-ISO-12185-1996-327309_SAIG_CEN_CEN_753930/.

García-Aparicio, M. P., Oliva, J. M., Manzanares, P., Ballesteros, M., Ballesteros, I., González, A., & Negro, M. J. (2011). Second-generation ethanol production from steam exploded barley straw by Kluyveromyces marxianus CECT 10875. *Fuel, 90*(4), 1624–1630. https://doi.org/10.1016/j.fuel.2010.10.052.

Gautam, A., & Agarwal, A. K. (2015). Determination of important biodiesel properties based on fuel temperature correlations for application in a locomotive engine. *Fuel, 142*, 289–302. https://doi.org/10.1016/j.fuel.2014.10.032.

Geng, P., Cao, E., Tan, Q., & Wei, L. (2017). Effects of alternative fuels on the combustion characteristics and emission products from diesel engines: A review. *Renewable and Sustainable Energy Reviews, 71*, 523–534. https://doi.org/10.1016/j.rser.2016.12.080.

Ghimire, A., Pirozzi, F., Trably, E., Escudie, R., Lens, P. N. L., & Esposito, G. (2015). A review on dark fermentative biohydrogen production from organic biomass: Process parameters and use of by-products. *Applied Energy, 144*, 73–95. https://doi.org/10.1016/j.apenergy.2015.01.045.

Gottumukkala, L. D., Parameswaran, B., Valappil, S. K., Mathiyazhakan, K., Pandey, A., & Sukumaran, R. K. (2013). Biobutanol production from rice straw by a non acetone producing Clostridium sporogenes BE01. *Bioresource Technology, 145*, 182–187. https://doi.org/10.1016/j.biortech.2013.01.046.

Gupta, A., & Verma, J. P. (2015). Sustainable bio-ethanol production from agro-residues: A review. *Renewable and Sustainable Energy Reviews, 41*, 550–567. https://doi.org/10.1016/j.rser.2014.08.032.

Guo, X. M., Trably, E., Latrille, E., Carrre, H., & Steyer, J. P. (2010). Hydrogen production from agricultural waste by dark fermentation: A review. *International Journal of Hydrogen Energy, 35*(19), 10660–10673. https://doi.org/10.1016/j.ijhydene.2010.03.008.

Gupta, R. B., & Demirbas, A. (2010). Gasoline, diesel and ethanol biofuels from grasses and plants. In R. B. Gupta (Ed.), *Gasoline, diesel and ethanol biofuels from grasses and plants*. Cambridge University Press.

Han, M., Kang, K. E., Kim, Y., & Choi, G. W. (2013). High efficiency bioethanol production from barley straw using a continuous pretreatment reactor. *Process Biochemistry, 48*(3), 488–495. https://doi.org/10.1016/j.procbio.2013.01.007.

He, T., Chen, Z., Zhu, L., & Zhang, Q. (2018). The influence of alcohol additives and EGR on the combustion and emission characteristics of diesel engine under high-load condition. *Applied Thermal Engineering, 140*, 363–372. https://doi.org/10.1016/j.applthermaleng.2018.05.064.

Hijosa-Valsero, M., Paniagua-García, A. I., & Díez-Antolínez, R. (2017). Biobutanol production from apple pomace: The importance of pretreatment methods on the fermentability of lignocellulosic agro-food wastes. *Applied Microbiology and Biotechnology, 101*(21), 8041–8052. https://doi.org/10.1007/s00253-017-8522-z.

Hu, B. Bin, Li, M. Y., Wang, Y. T., & Zhu, M. J. (2018). High-yield biohydrogen production from non-detoxified sugarcane bagasse: Fermentation strategy and mechanism. *Chemical Engineering Journal, 335*, 979–987. https://doi.org/10.1016/j.cej.2017.10.157.

Ibrahim, M. F., Abd-Aziz, S., Razak, M. N. A., Phang, L. Y., & Hassan, M. A. (2012). Oil palm empty fruit bunch as alternative substrate for acetone-butanol-ethanol production by Clostridium butyricum EB6. *Applied Biochemistry and Biotechnology, 166*, 1615–1625. https://doi.org/10.1007/s12010-012-9538-6.

International Energy Agency. (2018). Global energy & CO_2 status report. In IEA (Ed.), *Global energy & CO_2 status report*. International Energy Agency.

Joshi, S. M., Waghmare, J. S., Sonawane, K. D., & Waghmare, S. R. (2015). Bio-ethanol and bio-butanol production from orange peel waste. *Biofuels, 6*(2), 55–61. https://doi.org/10.1080/17597269.2015.1045276.

Karmakar, B., & Halder, G. (2019). Progress and future of biodiesel synthesis: Advancements in oil extraction and conversion technologies. *Energy Conversion and Management, 182*, 307–339. https://doi.org/10.1016/j.enconman.2018.12.066.

Kattimani, V. R., Venkatesha, B. M., & Ananda, S. (2014). Biodiesel production from unrefined rice bran oil through three-stage transesterification. *Advances in Chemical Engineering and Science, 4*(3), 361–366. https://doi.org/10.4236/aces.2014.43039.

Kim, S., & Kim, C. H. (2013). Bioethanol production using the sequential acid/alkali-pretreated empty palm fruit bunch fiber. *Renewable Energy, 54*, 150–155. https://doi.org/10.1016/j.renene.2012.08.032.

Kong, X., Xu, H., Wu, H., Wang, C., He, A., Ma, J., … Ouyang, P. (2016). Biobutanol production from sugarcane bagasse hydrolysate generated with the assistance of gamma-valerolactone. *Process Biochemistry*. https://doi.org/10.1016/j.procbio.2016.06.013.

Kuang, Y., Zhang, Y., Zhou, B., Li, C., Cao, Y., Li, L., & Zeng, L. (2016). A review of renewable energy utilization in islands. *Renewable and Sustainable Energy Reviews, 59*, 504–513. https://doi.org/10.1016/j.rser.2016.01.014.

Li, K., Liu, R., & Sun, C. (2016). A review of methane production from agricultural residues in China. *Renewable and Sustainable Energy Reviews, 54*, 857–865. https://doi.org/10.1016/j.rser.2015.10.103.

Li, X., Dang, F., Zhang, Y., Zou, D., & Yuan, H. (2015). Anaerobic digestion performance and mechanism of ammoniation pretreatment of corn stover. *BioResources*. https://doi.org/10.15376/biores.10.3.5777-5790.

Liew, L. N., Shi, J., & Li, Y. (2012). Methane production from solid-state anaerobic digestion of lignocellulosic biomass. *Biomass and Bioenergy*. https://doi.org/10.1016/j.biombioe.2012.09.014.

Li, Y., Tang, W., Chen, Y., Liu, J., Lee, C., & fon, F. (2019). Potential of acetone-butanol-ethanol (ABE) as a biofuel. *Fuel, 242*, 673–686. https://doi.org/10.1016/j.fuel.2019.01.063.

Lau, M. W., & Dale, B. E. (2009). Cellulosic ethanol production from AFEX-treated corn stover using *Saccharomyces cerevisiae* 424A(LNH-ST). *Proceedings of the National Academy of Sciences of the United States of America, 106*(5), 1368–1373. https://doi.org/10.1073/pnas.0812364106.

Lin, L., Ying, D., Chaitep, S., & Vittayapadung, S. (2009). Biodiesel production from crude rice bran oil and properties as fuel. *Applied Energy, 86*(5), 681–688. https://doi.org/10.1016/j.apenergy.2008.06.002.

Matzen, M., & Demirel, Y. (2016). Methanol and dimethyl ether from renewable hydrogen and carbon dioxide: Alternative fuels production and life-cycle assessment. *Journal of Cleaner Production, 139*, 1068–1077. https://doi.org/10.1016/j.jclepro.2016.08.163.

Mendonça, I. M., Paes, O. A. R. L., Maia, P. J. S., Souza, M. P., Almeida, R. A., Silva, C. C., … de Freitas, F. A. (2019). New heterogeneous catalyst for biodiesel production from waste tucumã peels (Astrocaryum aculeatum Meyer): Parameters optimization study. *Renewable Energy*. https://doi.org/10.1016/j.renene.2018.06.059.

Munari, B. F., Cavagnino, D., & Scientific, T. F. (2009). Determination of total FAME and linolenic acid methyl ester in pure biodiesel (B100) by GC in compliance with EN 14103. *Test*.

Nasr, N., Gupta, M., Elbeshbishy, E., Hafez, H., El Naggar, M. H., & Nakhla, G. (2014). Biohydrogen production from pretreated corn cobs. *International Journal of Hydrogen Energy, 39*(75), 19921–19927. https://doi.org/10.1016/j.ijhydene.2014.10.004.

Noomtim, P., & Cheirsilp, B. (2011). Production of butanol from palm empty fruit bunches hydrolyzate by Clostridium acetobutylicum. *Energy Procedia, 9*, 140–146. https://doi.org/10.1016/j.egypro.2011.09.015.

O-Thong, S., Boe, K., & Angelidaki, I. (2012). Thermophilic anaerobic co-digestion of oil palm empty fruit bunches with palm oil mill effluent for efficient biogas production. *Applied Energy.* https://doi.org/10.1016/j.apenergy.2011.12.092.

Onu, P., Abolarin, M. S., & Anafi, F. O. (2017). Assessment of effect of rice husk ash on burnt properties of badeggi clay. *International Journal of Advanced Research, 5*(5), 240–247. https://doi.org/10.21474/IJAR01/4103.

Onu, P., & Mbohwa, C. (2018). Correlation between future energy systems and industrial revolutions. *Proceedings of the international conference on industrial engineering and operations management*, 1953–1961 (Pretoria/Johannesburg).

Onu, P., & Mbohwa, C. (2019a). Sustainable production: New thinking for SMEs. *Journal of Physics: Conference Series, 1378*(2). https://doi.org/10.1088/1742-6596/1378/2/022072.

Onu, P., & Mbohwa, C. (2019b). Sustainable supply chain management: Impact of practice on manufacturing and industry development. *Journal of Physics: Conference Series, 1378*(2). https://doi.org/10.1088/1742-6596/1378/2/022073.

Onu, P., & Mbohwa, C. (2019c). Renewable energy technologies in brief. *International Journal of Scientific and Technology Research, 8*(10), 1283–1289.

Onu, P., & Mbohwa, C. (2019d). Industrial energy conservation initiative and prospect for sustainable manufacturing. *Procedia Manufacturing, 35*, 546–551. https://doi.org/10.1016/j.promfg.2019.05.077.

Pang, Z. W., Lu, W., Zhang, H., Liang, Z. W., Liang, J. J., Du, L. W., … Feng, J. X. (2016). Butanol production employing fed-batch fermentation by Clostridium acetobutylicum GX01 using alkali-pretreated sugarcane bagasse hydrolysed by enzymes from Thermoascus aurantiacus QS 7-2-4. *Bioresource Technology.* https://doi.org/10.1016/j.biortech.2016.04.013.

Papargyropoulou, E., Lozano, R., Steinberger, J. K., Wright, N., & Ujang, Z. Bin (2014). The food waste hierarchy as a framework for the management of food surplus and food waste. *Journal of Cleaner Production, 76*, 106–115. https://doi.org/10.1016/j.jclepro.2014.04.020.

Patil, S. S., Kar, A., & Mohapatra, D. (2016). Stabilization of rice bran using microwave: Process optimization and storage studies. *Food and Bioproducts Processing, 99*, 204–211. https://doi.org/10.1016/j.fbp.2016.05.002.

Paul, S., & Dutta, A. (2018). Challenges and opportunities of lignocellulosic biomass for anaerobic digestion. *Resources, Conservation and Recycling, 130*, 164–174. https://doi.org/10.1016/j.resconrec.2017.12.005.

Pereira, L. G., Dias, M. O. S., Mariano, A. P., Maciel Filho, R., & Bonomi, A. (2015). Economic and environmental assessment of n-butanol production in an integrated first and second generation sugarcane biorefinery: Fermentative versus catalytic routes. *Applied Energy, 160*, 120–131. https://doi.org/10.1016/j.apenergy.2015.09.063.

Procentese, A., Raganati, F., Olivieri, G., Elena Russo, M., & Marzocchella, A. (2017). Pretreatment and enzymatic hydrolysis of lettuce residues as feedstock for bio-butanol production. *Biomass and Bioenergy, 96*, 172–179. https://doi.org/10.1016/j.biombioe.2016.11.015.

Protocol, F. L. W. (2016). *Food loss and waste accounting and reporting standard.* FLW Protocol.

Qiu, J., Ma, L., Shen, F., Yang, G., Zhang, Y., Deng, S., ... Hu, Y. (2017). Pretreating wheat straw by phosphoric acid plus hydrogen peroxide for enzymatic saccharification and ethanol production at high solid loading. *Bioresource Technology*. https://doi.org/10.1016/j.biortech.2017.04.040.

Qureshi, N., Cotta, M. A., & Saha, B. C. (2014). Bioconversion of barley straw and corn stover to butanol (a biofuel) in integrated fermentation and simultaneous product recovery bioreactors. *Food and Bioproducts Processing, 92*(3), 298–308. https://doi.org/10.1016/j.fbp.2013.11.005.

Ren, H., Mei, Z., Fan, W., Wang, Y., Liu, F., Luo, T., ... Feng, R. (2018). Effects of temperature on the performance of anaerobic co-digestion of vegetable waste and swine manure. *International Journal of Agricultural and Biological Engineering*. https://doi.org/10.25165/j.ijabe.20181101.3706.

Saha, B. C., & Cotta, M. A. (2010). Comparison of pretreatment strategies for enzymatic saccharification and fermentation of barley straw to ethanol. *New Biotech, 27*(1). https://doi.org/10.1016/j.nbt.2009.10.005.

Saini, J. K., Saini, R., & Tewari, L. (2015). Lignocellulosic agriculture wastes as biomass feedstocks for second-generation bioethanol production: Concepts and recent developments. *Biotech, 3*(5), 337–353. https://doi.org/10.1007/s13205-014-0246-5.

Sakuragi, K., Li, P., Otaka, M., & Makino, H. (2016). Recovery of bio-oil from industrial food waste by liquefied dimethyl ether for biodiesel production. *Energies, 9*(2). https://doi.org/10.3390/en9020106.

Sangyoka, S., Reungsang, A., & Lin, C. Y. (2016). Optimization of biohydrogen production from sugarcane bagasse by mixed cultures using a statistical method. *Sustainable Environment Research, 26*(5), 235–242. https://doi.org/10.1016/j.serj.2016.05.001.

Scano, E. A., Asquer, C., Pistis, A., Ortu, L., Demontis, V., & Cocco, D. (2014). Biogas from anaerobic digestion of fruit and vegetable wastes: Experimental results on pilot-scale and preliminary performance evaluation of a full-scale power plant. *Energy Conversion and Management, 77*, 22–30. https://doi.org/10.1016/j.enconman.2013.09.004.

Sgroi, F., Donia, E., & Alesi, D. R. (2018). Renewable energies, business models and local growth. *Land Use Policy, 72*(September 2017), 110–115. https://doi.org/10.1016/j.landusepol.2017.12.028.

Shahbaz, M., Ammar, M., Zou, D., Korai, R. M., & Li, X. J. (2019). An insight into the anaerobic Co-digestion of municipal solid waste and food waste: Influence of Co-substrate mixture ratio and substrate to inoculum ratio on biogas production. *Applied Biochemistry and Biotechnology, 187*(4), 1356–1370. https://doi.org/10.1007/s12010-018-2891-3.

Sharma, Y. C., & Singh, B. (2010). A hybrid feedstock for a very efficient preparation of biodiesel. *Fuel Processing Technology, 91*(10), 1267–1273. https://doi.org/10.1016/j.fuproc.2010.04.008.

Shedrack, G. M., Yawas, D. S., Emmanuel, S. G., Onu, P., Dagwa, I. M., & Ibrahim, G. Z. (2019). *Effect of postweld heat treatment on the mechanical behaviour of austinitic stainless steel, 6* pp. 1–9). https://doi.org/10.5281/zenodo.2639252 (4).

Shi, X., Lin, J., Zuo, J., Li, P., Li, X., & Guo, X. (2017). Effects of free ammonia on volatile fatty acid accumulation and process performance in the anaerobic digestion of two typical bio-wastes. *Journal of Environmental Sciences, 55*, 49–57. https://doi.org/10.1016/j.jes.2016.07.006.

Smith, A. D., Holtzapple, M. T., Janke, L., Leite, A., Batista, K., Weinrich, S., ... White, S. (2015). Impact of the substrate loading regime and phosphoric acid supplementation on performance of biogas reactors and microbial community dynamics during anaerobic digestion of chicken wastes. *Bioresource Technology.* https://doi.org/10.1016/j.biortech.2010.01.083.

Stephen, J. L., & Periyasamy, B. (2018). Innovative developments in biofuels production from organic waste materials: A review. *Fuel, 214*, 623−633. https://doi.org/10.1016/j.fuel.2017.11.042.

Szulczyk, K. R. (2010). Which is a better transportation fuel−butanol or ethanol? *International Journal of Energy and Environment.*

Talebnia, F., Karakashev, D., & Angelidaki, I. (2010). Production of bioethanol from wheat straw: An overview on pretreatment, hydrolysis and fermentation. *Bioresource Technology, 101*, 4744−4753. https://doi.org/10.1016/j.biortech.2009.11.080.

Tandon, M., Thakur, V., Tiwari, K. L., & Jadhav, S. K. (2018). Enterobacter ludwigii strain IF2SW-B4 isolated for bio-hydrogen production from rice bran and de-oiled rice bran. *Environmental Technology and Innovation, 10*, 345−354. https://doi.org/10.1016/j.eti.2018.03.008.

Taslim, Iriany, Bani, O., Parinduri, S. Z. D. M., & Ningsih, P. R. W. (2018). Biodiesel production from rice bran oil by transesterification using heterogeneous catalyst natural zeolite modified with K_2CO_3. *IOP Conference Series: Materials Science and Engineering.* Article 012107. https://doi.org/10.1088/1757-899X/309/1/012107.

Tie, S. F., & Tan, C. W. (2013). A review of energy sources and energy management system in electric vehicles. *Renewable and Sustainable Energy Reviews.* https://doi.org/10.1016/j.rser.2012.11.077.

Urbaniec, K., & Bakker, R. R. (2015). Biomass residues as raw material for dark hydrogen fermentation - a review. *International Journal of Hydrogen Energy, 40*(9), 3648−3658. https://doi.org/10.1016/j.ijhydene.2015.01.073.

Vallinayagam, R., Vedharaj, S., Yang, W. M., Roberts, W. L., & Dibble, R. W. (2015). Feasibility of using less viscous and lower cetane (LVLC) fuels in a diesel engine: A review. *Renewable and Sustainable Energy Reviews, 51*, 1166−1190. https://doi.org/10.1016/j.rser.2015.07.042.

Wang, Z., Chen, X., Huang, S., Chen, Y., MacK, J. H., Tang, J., … Huang, R. (2016). Visualization study for the effects of oxygen concentration on combustion characteristics of water-emulsified diesel. *Fuel.* https://doi.org/10.1016/j.fuel.2016.03.010.

Wang, H., Wang, J., Fang, Z., Wang, X., & Bu, H. (2010). Enhanced bio-hydrogen production by anaerobic fermentation of apple pomace with enzyme hydrolysis. *International Journal of Hydrogen Energy, 35*(15), 8303−8309. https://doi.org/10.1016/j.ijhydene.2009.12.012.

Wang, X., Lu, X., Li, F., & Yang, G. (2014). Effects of temperature and Carbon-Nitrogen (C/N) ratio on the performance of anaerobic co-digestion of dairy manure, chicken manure and rice straw: Focusing on ammonia inhibition. *PloS One.* https://doi.org/10.1371/journal.pone.0097265.

Wikandari, R., Nguyen, H., Millati, R., Niklasson, C., & Taherzadeh, M. J. (2015). Improvement of biogas production from orange peel waste by leaching of limonene. *BioMed Research International.* Article 94182. https://doi.org/10.1155/2015/494182.

Wu, X., Yao, W., & Zhu, J. (2010). Effect of pH on continuous biohydrogen production from liquid swine manure with glucose supplement using an anaerobic sequencing batch reactor. *International Journal of Hydrogen Energy, 35*(13), 6592−6599. https://doi.org/10.1016/j.ijhydene.2010.03.097.

Yin, D., Liu, W., Zhai, N., Yang, G., Wang, X., Feng, Y., & Ren, G. (2014). Anaerobic digestion of pig and dairy manure under photo-dark fermentation condition. *Bioresource Technology, 166*, 373−380. https://doi.org/10.1016/j.biortech.2014.05.037.

Yuan, W., Gong, Z., Wang, G., Zhou, W., Liu, Y., Wang, X., & Zhao, M. (2018). Alkaline organosolv pretreatment of corn stover for enhancing the enzymatic digestibility. *Bioresource Technology, 265*, 464–470. https://doi.org/10.1016/j.biortech.2018.06.038.

Zeshan, Karthikeyan, O. P., & Visvanathan, C. (2012). Effect of C/N ratio and ammonia-N accumulation in a pilot-scale thermophilic dry anaerobic digester. *Bioresource Technology, 113*, 294–302. https://doi.org/10.1016/j.biortech.2012.02.028.

Zhang, T., Jacobson, L., Björkholtz, C., Munch, K., & Denbratt, I. (2016). Effect of using butanol and octanol isomers on engine performance of steady state and cold start ability in different types of diesel engines. *Fuel, 184*, 708–717. https://doi.org/10.1016/j.fuel.2016.07.046.

Zhang, Y., Wong, W. T., & Yung, K. F. (2013). One-step production of biodiesel from rice bran oil catalyzed by chlorosulfonic acid modified zirconia via simultaneous esterification and transesterification. *Bioresource Technology, 147*, 59–64. https://doi.org/10.1016/j.biortech.2013.07.152.

Zhao, Y., Damgaard, A., & Christensen, T. H. (2018). Bioethanol from corn stover – a review and technical assessment of alternative biotechnologies. *Progress in Energy and Combustion Science, 67*, 275–291. https://doi.org/10.1016/j.pecs.2018.03.004.

Zhou, N., Huo, M., Wu, H., Nithyanandan, K., Lee, C., fon, F., & Wang, Q. (2014). Low temperature spray combustion of acetone-butanol-ethanol (ABE) and diesel blends. *Applied Energy, 117*, 104–115. https://doi.org/10.1016/j.apenergy.2013.11.035.

CHAPTER

7 Overview of models for agricultural waste management, and trends in biofuels production

7.1 Introduction

The use of fossil fuels would soon become a thing of the past due to the unfavorable and adverse effects, which do not support sustainable ecosystems (Abas, Kalair, & Khan, 2015; Bhowmik, Bhowmik, Ray, & Pandey, 2017). As such, it has ushered different scientific interventions and innovations to implement viable systems for energy recovery from cost-effective processes and agrowaste/biomasses in the future ("Innov. Strateg. Food Ind.," 2016). The release and generation of wastes have continued to rise due to the growing global population, which pressures life forms and impacts negatively on the ecosystem and natural habitat. This has led to the search for nascent pathways for energy generation from agricultural wastes, which now finds global acceptance and application (Gupta & Demirbas, 2010; Onu & Mbohwa, 2018b; Welfle, Gilbert, Thornley, & Stephenson, 2017a, 2017b). Energy generation in the form of fuels (solid and liquid) for transportation, heat, or electricity is processed through different mechanisms and technological advances called waste to energy (WTE). The WTE initiative is essential to establish a balance between waste management and energy security, while also considers policies that guide the exploitation of agricultural and food-based produce, technology in use, and energy production specifications (Brunner & Rechberger, 2015; Demirbas, 2011).

Although the use of fossil fuel is found to be vital in numerous modern applications, from domestic to industrial or as a source of income for many developing and underdeveloped nations, this energy source contributes to not less than 80% of the global energy requirement. It has the potential to be used continually for decades to come due to its abundance (Onu & Mbohwa, 2018a). Regardless, the awareness that the world we live in is faced with severe climate degradation effects, and setbacks, caused by anthropogenic engagements, bring about the prospect of a green society and zero-waste disposition. The latter considers the route to sustainable development (advancement of WTE initiatives) through waste recovery for commercially viable clean and renewable energy supply (Onu & Mbohwa, 2019c). The potential of biodegradable, agricultural wastes is tenable for the replacement of

Agricultural Waste Diversity and Sustainability Issues. https://doi.org/10.1016/B978-0-323-85402-3.00003-6

tons of minerals and fossil-based fuels, used for generating electricity. These resolves are achieved through controlled burning for substitution of coal, resulting in minimal emission of pollutants and overcoming the challenges due to waste disposal leading to the abandonment of useful lands. Moreover, the agrowastes such as ashes have potential in the development of composite, offering good mechanical properties in cement production (Lawal et al., 2019; Onu, Abolarin, & Anafi, 2017), thus eliminating the challenge of methane release to the atmosphere.

The different techniques for waste management, with regard to several essential considerations, include technology type, governmental involvement in policies, decision implementation, raw material availability, and skill/technology maturity to mention a few (Babayemi, Ogundiran, & Osibanjo, 2017; Davis et al., 2016; Meena et al., 2019; Onu & Mbohwa, 2019f). However, the certainty for the deployment of the different interventions will lead the various waste management initiatives for gainful potentials, and the WTE campaign requires precise planning due to the complexity and essence of the operations involved. A case study assessment of agricultural waste management success has been addressed through the perspective of mathematical modeling to improve the operational processes involved while applying the sustainability values (Edalatpour, Al-e-hashem, Karimi, & Bahli, 2018). This chapter explores effective wastes management practices with an overview of the aforementioned factors. A descriptive application of a linear and mixed-integer mathematical model is analyzed to determine the success of the minimal cost of waste management services.

The trends noticed, among developing countries, show the inability and lack of technical capabilities to properly handle wastes through the traditional procedures from treatments and biobeneficiation for economic and environmental valuation. The availability of biomasses, from viable plants (of the different generation of energy feedstocks) and animal wastes, is a vital condition for WTE actualization. It has been established that the biodigester technology is capable of producing sufficient electricity from methane gas collected during the digestive process and, thus, saves the environment from the discomfort of poorly managed biowaste landfills. Also, the controlled burning of biowastes for energy disposition contributes to the reduction of carbon dioxide and climate change mitigation (Walter Borges de Oliveira, Leoneti, Magrini Caldo, & Borges de Oliveira, 2011). Notably, instead of having wastes transported to landfills, the wastes can be transformed to generate energy, leaving very little substrates that can still be reused (manure, reinforcement material, other industrial usages), thereby saving lands and contributing to the ecosystem safety.

The focus, therefore, must be to develop models and techniques that will support the improvement in agrowaste management operations, to promote waste estimation, segregation, quantification, collection and analysis, reuse, and auditing and report. Numerous self-acclaimed approaches exist for these wastes' evaluations (Yusuf et al., 2019). However, more studies are expected to revolutionize the idea. The emergence and proliferation of information communication technologies and the advent of GIS, extensive area networks, drones, and sensor-based technologies among others are fast transforming the waste management conception (Abdel-

Shafy & Mansour, 2018; Onu & Mbohwa, 2019d). The motivation here is to find means to implement excellent operational tactics through different measures that will guide decisions in agrowaste management and the systems thereof. A look at the conversion of the bioresidues to clean renewable energies for sustainability improvement, putting into consideration the technological perspectives and models for cost-effective management systems, is presented in the next section.

7.2 Emerging opportunities, challenges, and prospects of modern bioenergy process

The euphoria concerning WTE strategies that operate with minimum emission and maximum energy yield is dependent on the technology that is used (Lega, D'Antonio, & Napoli, 2010; Ross, 2018). Separate from the biochemical and thermochemical processes to achieve WTE, other emerging technology prospects exist; these include microbial electrolysis cell (MEC) and microbial fuel cell (MFC). The expectations based on the effective operation of any chosen techniques present new expectations for performance measure with regard to minimum temperature and high output energy. The two most popular routes to achieve WTE include the biochemical and thermochemical processes. The biochemical process considers anaerobic conversion and fermentation of biodegradable biomasses for the production of biogases. The thermochemical approach includes the incineration, pyrolysis, and gasification processes (Khan et al., 2017; Santoro, Arbizzani, Erable, & Ieropoulos, 2017). The thermochemical procedures are characterized by different features and requirements, which make one option more preferable for wastes remediation over the other. Table 7.1 describes the framework for thermal conversion. Also, the subsequent section (Techniques in Biodegradable Waste Conversion Process) highlights an essential consideration for WTE prospect through different technological approaches that are applicable to actualize solid waste management operation.

Biological conversion technique

Anaerobic digestion process

Anaerobic digestion or methanogenesis is a process where microbial activities facilitate the degradation and breakdown of organic, biodegradable compounds in the absence of oxygen to produce biogas (methane) and potent slurry, which can further be treated and used for other purposes (e.g., plant vases). The slurry can also be adapted for the nourishment of soil (Ezejiofor, 2014; Tambone, Genevini, D'Imporzano, & Adani, 2009; Vaish et al., 2019). The conversion phases are as follows: hydrolysis (carbohydrates, fat, fat, and protein present), acedogenesis (fatty acid, sugar, and amino acid present), acetogenesis (a form of volatile fatty acid), and the final stage—methanogenesis (methane, carbon dioxide, and other trace gases). The digestion process is essential to meet cleaner energy production, environmental pollution, wastes volume reduction, and recycling of nutrients (Liebetrau, Sträuber,

Table 7.1 Features of thermochemical process.

Approach	Parameters						
	Operation principle	Operating condition	Operating temperature (°C)	Pretreatment requirement	Product yield		
					Solid	Liquid	Gas
Incinerator	Full oxidative combustion	Presence of sufficient oxygen	850–1200	Not necessary	Bottom ash, fly ash, slag, other noncombustible substances such as metals and glass	–	CO_2, H_2O, O_2, N_2
Pyrolysis	Thermal degradation of organic materials in the absence of oxygen	Absence of oxygen	400–800	Required	Ash, char (the combination of noncombustibles and carbon)	Condensate of pyrolysis gas (pyrolysis oil, wax, tar)	–
Gasificat on	Partial oxidation	Controlled supply of oxygen	800–1600	Required	Ash, slag	–	Syngas (H_2, CO, CO_2, CH_4, H_2O, N_2)

Kretzschmar, Denysenko, & Nelles, 2019; Pham, Kaushik, Parshetti, Mahmood, & Balasubramanian, 2015). Certain challenges that impede the success/proliferation of biodigestion include a lengthy period and waiting time, which lasts up to 40 days. The characteristics of the feedstock, based on the occurrence and the percentage composition of carbon to nitrogen (C: N), contained in the biomass substrate, being fed into the digester (Lyu, Shao, Akinyemi, & Whitman, 2018). As such, a higher proportion of protein-based feedstock, leading to the production of ammonia, will cause an imbalance in the digestion process and reduce the rate of methanogenesis (Fotidis, Karakashev, Kotsopoulos, Martzopoulos, & Angelidaki, 2013), whereas the reverse is the case for abundant carbohydrates for a sustainable supply of biogas (Steinbusch, Arvaniti, Hamelers, & Buisman, 2009).

Fermentation process

The process where the production of biohydrogen or ethanol occurs is termed fermentation. The system is such that organic wastes, mixed in a fluid, are acted upon by different microbes so that carbohydrates are converted to sugar. The hydrolyzed sugar is acted upon by bacteria (*Clostridium* and *Enterobacter*) responsible for the production of hydrogen and short-chain fatty acids. Recently, more research focused on clean energy is directed toward the production of hydrogen of high energy yield (122 kg/g) and preferable, compared with most hydrocarbon fuels (Schlüter et al., 2008). Moreover, the application of hydrogen fuels in combustion engines produces little or no harmful emissions and promotes climate change mitigation. One of the most promising and emerging technologies for biohydrogen production is through the dart fermentation process. In this process, the production yields can be increased to a theoretical value of 4 mol/mol of glucose according to Zong, Yu, Zhang, Fan, & Zhou (2009). However, this value appears to be lower than the actual yield, which, as such, is said to be constrained by the sensitivity of the bacteria strains, affected by different operational conditions, such as pH, hydrogen accumulation, and the presence of short-chain fatty acids.

Thermochemical conversion process

Unlike the biochemical conversion process, this technique is a dry process, requiring nonbiodegradable wastes, however, of green origin, to undergo thermal conversion for sustainable end use (cost-effective, eco-friendly fuel). The pretreatment operation in thermochemical conversion ensures the removal of noncombustible, wet items of the right size, under normal conditions (D. Zhang, Luo, Li, Wang, & Li, 2018). The process can facilitate value creation in energy and chemical production from wastes and dried agricultural discards (wood, husks, fruit backs, etc). As such, the process is completed in the absence of limited or no oxygen at elevated temperature to yield char, hydrocarbon oils, and gases as eco-friendly fuels (Table 7.1). The process also recovers energy from nonbiodegradable waste, which is a waste of energy advancement through the same thermal process, which in the long run contributes to the general reduction of waste (Zhang et al., 2018). The feedstocks are collected in a large compartment where high-temperature heating takes place. Heat is recovered from the process and used or channeled to generate steam power (turbine) for electricity generation.

Incineration

The terminology "incineration" is the measure of decomposition of waste at high temperatures. It is considered as one of the most effective thermal conversion approaches for waste beneficiation. As seen in Table 7.1, incineration processes do not require cumbersome pretreatment operations like other techniques. Hence, the waste resources free from unwanted obstacles can directly be fed into a heating compartment, according to the design, so that the air is introduced to facilitate the combustion. This results in very high temperatures due to the decomposition of the feedstock. As such, layers of gas fuels are formed at the grate, which travels upward, and trapped so that the heat potential can be utilized or converted to drive/power other mechanisms. The application of incineration technology is suggested for handling biomedical wastes disposal.

Wastes radioactive and highly hazardous pharmaceutical wastes are transformed into energy without leaving any causative effect on humans or their environment (Tan, Hashim, Lee, Taib, & Yan, 2014). Residues and by-products from incineration (bottom ash) can also be processed or converted to materials for road and building construction and soil amendment before resorting to the last option of disposal via landfill. Research has identified relative benefits of incineration method, ranging from minimized transportation of wastes, as a source replacement for fossil fuel combustion: reducing the burden of landfill activities, helping to eliminate methane gas emission, means for heat and power recovery, less land required compared with landfill sites, and reducing waste volume. The fluidized bed system type of incineration process converts solid wastes to liquid form and then to sustainable/clean fuel.

Pyrolysis

The theory behind the thermal treatment of waste is based on the restricted control of oxygen supply to control the burning and monitoring of the temperature at which decomposition occurs. Pyrolysis, unlike the incineration and gasification process, takes place in the absence of oxygen (Table 7.1), producing different proportions of gases, vapor, and char. Modeling of the process design to improve the pyrolytic technique has also been conducted (Ghasemi & Yusuff, 2016). The technology has also been researched to favorably handle plastic waste (Kirtania & Bhattacharya, 2012, 2015; Sharma, Pareek, & Zhang, 2015). It is a slightly ineffective process, allowing the conversion of about 70%, compared with the incineration technique where up to 90% of the feedstock can be decomposed during the combustion. Although, from popular saying "no waste is a waste," as such, the residues from this process can be transported to nearby farms for soil enrichment.

Gasification

The presence of water molecules in biomass feedstocks allows for the gasification of its residue, which is directly converted to syngas, described in Table 7.1, thus leading to the separation of organic and inorganic compounds (supercritical water gasification process). In this process, the conversion of feedstock is possible under intense heating, up to 1600°C, without the use of a catalyst. Researchers (Al-Salem, Lettieri, & Baeyens, 2009; Miandad, Barakat, Aburiazaiza, Rehan, & Nizami, 2016), have sorted means to produce sustainable fuel from biomasses without leaving any wastes (char). The development of the gasification technology is dependent on reactor design, categorized accordingly: moving or fixed bed, entrained flow, and the fluidized bed gasifier, and this has been extensively researched through time (Peterson et al., 2008). However, it is worth noting that the pyrolysis and gasification conversion technology/systems are still within their experimentation stages and have not yet attained technological maturity for commercialization. Several considerations have led to the aforementioned resolve: not enough data and operational strategy to develop the technologies beyond the pilot/experimental stages.

7.3 Model case scenarios for agrowaste management system

Nowadays, agitation for cleaner energy production applies different strategic approaches, models (mathematical), and systems (technologies) that can effectively transform the waste management operations to lead operational effectiveness (Heidenreich & Foscolo, 2015; Onu & Mbohwa, 2019b, 2019a; Onu & Mbohwa, 2019e; Umeda, Nakamura, Lu, & Yoshikawa, 2019). Different waste management policies exist and vary from country to country. One of the most popular waste management practices is the "PAYT—pay as you throw" and the "deposit refund system" according to Yau (2010). However, the above-identified waste management approach seemed effective since it only considered the volume of waste generation as at the time of receiving the conception; moreover, there were gaps, as to their extension to meet future disposition and waste management protocols, including advanced communication and digitization of the processes involved for tracking and monitoring. Current development features the adoption of Web-GIS-based systems, which have been tested in advanced economies such as China and Germany, and supports the optimization of waste management operations (Reichenbach, 2008). Hence, the proposition of mathematical models assimilates data generation and information about future waste management practices.

This section demonstrates the development of a mathematical model for agrowaste management, which has received little attention in recent times (integer linear programming and mixed-integer linear programming). According to research, a model for municipal waste collection and routes for transporting it has been investigated (Dai, Li, & Huang, 2011; Di Bella, Di Trapani, Mannina, & Viviani, 2012; Ghiani, Laganà, Manni, Musmanno, & Vigo, 2014; Mavrotas, Skoulaxinou, Gakis, Katsouros,

& Georgopoulou, 2013). Also, the use of a mixed-integer programming model has been adopted for solid municipal wastes (Das & Bhattacharyya, 2015). Accordingly (Badran & El-Haggar, 2006; Fiorucci, Minciardi, Robba, & Sacile, 2003), researchers demonstrated the use of mixed and linear and integer programming techniques for waste management in municipalities and cities, respectively, with the intent to better coordinate waste management operations. In this chapter, the authors build upon the approach used by Ghiani et al. (2014), applying practical constraints' that demonstrate the integer linear and mixed-integer model for agrowaste management.

Application of mathematical modeling for agrowaste management

The model propositions for agrowaste management system discussed in this chapter draw an inference from the perspective of agricultural wastes to mitigate costs while developing the decision process to promote clean energy production. The basis for investment is focused on transport logistics (number of transporting system and quantity moved) involved in the waste management process and from the context of a case study research conducted in Uganda, where the investigations were undertaken. The essential considerations and assumptions in the present case scenario for the agrowaste management system include the following lists: (1) the energy stations are closest and at the center of the agrowaste collection hot spots, (2) wastes classification is completed at the various hot spots, (3) operational assessment of waste management activities and the system are conducted daily, (4) truck maintenance and the use of standby truck are inevitable, occurring daily, and (5) analysis of the transportation costs is based on a round trip per day. Also, the network nodes represent, at any particular time, the waste collection hot spot, clean energy conversion station, farm, and the standby truck parking (space).

Integer linear programming

In this section, the use of integer linear programming is demonstrated as a mathematical model for future biodegradable waste management operations. Tables 7.2 and 7.3 describes the indices and variables required to organize an optimally functional and sustainable agrowaste management system. The four nodes are considered for the conceptualized model including a waste collection hot spot, biodigester station, truck parking/field/warehouse, and a potential farm, undergoing repair (soil amendment site) and manure for enriched farming activities. The route can lead from the waste collection hot spot to the biodigester station and farm. Also, the remains from the digester can be transported directly to the farming area. The standby trucks are

Table 7.2 Indices used in the mathematical model.

Indices	Description
$i = 1,2,\dots,I$	Location of waste collection hotspot
$j = 1,2,\dots,J$	Location of biodigester
$k = 1,2,\dots,K$	Location of farm/landfill
$w = 1,2,\dots,W$	Location of truck parking

Table 7.3 Variables used in the mathematical model.

Variables	Description
$\widetilde{X}_{ij}, \widetilde{X}_{ik}$	Number of travels from waste hot spot to biodigester and farms
$\widetilde{x}_{ij}, \widetilde{x}_{ik}$	Number of trucks required from waste hot spot to digester location and farms
$\widetilde{Y}_{jk}$	Number of travels from the digester to farms
$\widetilde{y}_{jk}$	Number of required travels from biodigester plant to farm
$\widetilde{n}_{wi}, \widetilde{n}_{wj}$	Number of trucks required on standby from parking to waste hot spot, digester
$\widetilde{z}_w$	Decision to rent parking for standby truck
$\widetilde{z}_j$	The decision to construct the digester
$\widetilde{z}_k$	Decision to avail/select the farm (cite landfill)
$\widetilde{w}_j$	Quantity of waste to the biodigester
$\widetilde{t}_k$	Quantity of waste to the farm (biodegradable), daily
$\widetilde{w}_k$	Quantity of all category of waste resources transported to the farm
T_n	Total number of operational trucks
T_a	Total number of trucks on standby

provided and can transport resources from the hot spots to the energy station (location of biodigester/plant), without heading to the farms since there is no resource flow from the farm. Table 7.4 illustrates essential variables for consideration in the mathematical model. The number of trips between specific nodes on a daily basis is represented by $\widetilde{X}_{ij}\widetilde{X}_{ih}$, and $\widetilde{Y}_{jk}$, whereas $\widetilde{x}_{ij}, \widetilde{x}_{ik}, \widetilde{y}_{jk}, \widetilde{n}_{wi}, \widetilde{n}_{wj}$ represent the total of the trucks per day that have transported either resources/waste to specific nodes. Furthermore, $\widetilde{z}_w, \widetilde{z}_j$, and $\widetilde{z}_k$, and $\widetilde{w}_j, \widetilde{t}_k$, and $\widetilde{w}_k$ represent a decision, whether to build/rent/lease the space for the standby truck parking, a digester, or the farming area, and the desired quantity of wastes, which can be delivered to any node in time, per day, respectively. Finally, the variables T_n and T_a represent the number of operational trucks and standby (in case of breakdown) truck used daily, respectively.

Table 7.4 refers to the data needed to solve the mathematical model. The data comprised the estimated number of trips before the next nodes cumulated on a daily basis, represented as $\widetilde{a}_{ij}, \widetilde{a}_{jk}$ and $\widetilde{b}_{jk}$; the costs of traveling between particular nodes per truck on daily basis are represented as $\widetilde{c}_{ij}, \widetilde{c}_{ik}$, and $\widetilde{d}_{jk}$. The costs of moving a truck on standby, from the parking to the point of collection of the agricultural and digester per trip, are represented as $\widetilde{e}_{wi}, \widetilde{e}_{wj}$. The measure of the slurry from the digester is denoted as $\boldsymbol{m}$, whereas the volume of the truck as the quantity of waste at the hot spot is i, d_i, and the revenue for biodigester per ton is represented as $\widetilde{c}_j$. The operating expenses of each node per ton are represented as $\widetilde{\gamma}_j$ and $\widetilde{\gamma}_k$, and the tendency of operational truck $\boldsymbol{p}$, to breakdown, along with the following conditions $\widetilde{Q}_i, \widetilde{Q}_j, \widetilde{Q}_k, \widetilde{Q}_w$, is all essential if the model is required to maximize waste resources. The unit-level cost of investment for the digester $\widetilde{\delta}_j$, farm operation $\widetilde{\delta}_k$, and parking of standby truck $\widetilde{\delta}_w$, or the unit-level cost of the f car is valued on their cost per day

Table 7.4 Description of the data requirement.

Data variables	Description
$\widetilde{a}_{ij}, \widetilde{a}_{ik}$	Number of travels from waste hot spot to biodigester and farm per truck per day
$\widetilde{b}_{jk}$	Number of travels from digester center to farm per each truck, daily
$\widetilde{c}_{ij}, \widetilde{c}_{ik}$	Transport cost from an agrowaste hot spot, digester, and farms per trip and truck, daily
$\widetilde{d}_{jk}$	Transport cost from the digester to farm per trip and truck, daily
$\widetilde{e}_{wi}, \widetilde{e}_{wj}$	Cost of operating standby truck from parking to waste hot spot, biodigester per trip
$\boldsymbol{m}$	Quantity of slurry from digester
α	The volume of the truck
d_i	Waste quantity, a source at $\boldsymbol{i}$
$\widetilde{c}_j$	Income from operating the digester per tonne
$\widetilde{\gamma}_j$	Operating expense of the digester per tonne
$\widetilde{\gamma}_k$	Operating expense of the farm operation per tonne
$\widetilde{\delta}_j, \widetilde{\delta}_k, \widetilde{\delta}_w$	The unit-level investment cost of the digester, farm operation, and standby truck parking
$\boldsymbol{f}$	Unit per truck
$\boldsymbol{p}$	Preventive maintenance work on an operational truck
$\widetilde{Q}_i, \widetilde{Q}_j, \widetilde{Q}_k, \widetilde{Q}_w$	The caring potential of each waste hot spot, biodigester, and farm as well as truck parking

basis. These are achieved when the capital cost on the project invested is divided by the life span (days).

The objective function of the integer linear programming model used is to ascertain optimal operating costs, as in the case of an agroindustrial waste management process. It considered the total management cost of handling wastes generated daily (F_1). This includes the cost of investment for the biodigester and parking for the standby trucks, the cost of the farm operation and the digester, and also cost of transportation to access either of the stations. The variable (F_2) represents cost due to movement of the standby truck to and from the agrowaste hot spots, and the digester plant (F_3) represents the cost of the truck. Lastly, (B) is used to represent the expected revenue from the biodigester operation.

$$F_1(z, w, X, Y) = \left[\sum_j \left(\widetilde{\delta}_j \widetilde{z}_j + \widetilde{\gamma}_j \widetilde{w}_j\right) + \sum_k \left(\widetilde{\delta}_k \widetilde{z}_k + \widetilde{\gamma}_k \widetilde{t}_k\right)\right] + \left[\sum_{ij} \left(\widetilde{c}_{ij} \alpha \widetilde{X}_{ij}\right) + \sum_{ik} \left(\widetilde{c}_{ik} \alpha \widetilde{X}_{ik}\right) + \sum_{jk} \left(\widetilde{d}_{jk} \alpha \widetilde{Y}_{jk}\right)\right] \tag{7.1}$$

$$F_2(n,z) = \sum_{wj} \tilde{e}_{wj}\tilde{n}_{wj} + \sum_{wi} \tilde{e}_{wi}\tilde{n}_{wi} + \sum_{w} \tilde{\delta}_w \tilde{z}_w \quad (7.2)$$

$$F_3(x,y,n) = \sum f(T_n + T_a) \quad (7.3)$$

$$B_{(w)} = \sum_j \tilde{c}_j (1-m)\tilde{w}_j \quad (7.4)$$

Hence,

$$F = F_1 + F_2 + F_3 - B \quad (7.5)$$

Four constraints are considered for the model, represented in Eqs. (7.6–7.9) to ensure zero-waste generation at individual waste collection hot spot. Constraint 7.6 is aimed at the essence of the most minimum waste generation at each site to be met; accordingly, it cannot be greater or equal to the original feedstock.

$$\sum_j \alpha \tilde{X}_{ij} + \sum_k \alpha \tilde{X}_{ik} \geq d_i, \quad i = 1, \ldots, I \quad (7.6)$$

Constraint 7.7 represents the quantity of residual biomass released to the farms from the biodigestion station, which should be lesser than the feed than the initial feedstock.

$$m\dot{w}_j \leq \sum_k \alpha \overrightarrow{Y}_{jk}, \quad j = 1, \ldots, J \quad (7.7)$$

Constraints 7.8 and 7.9 represent the quantity of waste transported to the biodigester station and farm, which must not exceed the volume of the wastes which the digester or farm had to manage.

$$\tilde{w}_j \leq \tilde{Q}_j \tilde{z}_j, \quad j = 1, 2, \ldots, J \quad (7.8)$$

$$\tilde{t}_k \leq \tilde{Q}_k \tilde{z}_k, \quad k = 1, 2, \ldots, k \quad (7.9)$$

Constraint 7.10 represents the expectation that the number of functional standby trucks should exceed those under maintenance, while constraint 7.11 represents that the number of operational vehicles should be more than the standby trucks.

$$\sum_{wj} \tilde{n}_{wj} + \sum_{wi} \tilde{n}_{wi} \geq pT_n \quad (7.10)$$

$$\sum_{w} \tilde{Q}_w \tilde{z}_w \leq T_n \quad (7.11)$$

Constraints 7.12–7.14 are indicative of the nonnegative variable integers. As such, they represent the number of trucks (standby trucks, not included) used for daily feedstock/waste transportation purpose.

$$\tilde{x}_{ij}, \text{integer} \geq 0, \; i(j) = 1, 2, \ldots, I, (J) \quad (7.12)$$

$$\tilde{x}_{ik}, \text{integer} \geq 0, \; i(k) = 1, 2, \ldots, I, (K) \quad (7.13)$$

$$\widetilde{y}_{jk}, \text{integer} \geq 0, \ j(k) = 1, 2, \ldots, J, (K) \tag{7.14}$$

Constraints 7.15 and 7.16 represent the nonnegative variable/integers, which show the number of standby trucks needed daily to execute proper waste management operation. The assumption with regard to the possible need to conduct maintenance/repair at the waste collection hot spot, digester station, or farm to cite the parking, for the standby truck:

$$\widetilde{n}_{wj}, \text{integer} \geq 0, \quad w(j) = 1, 2, \ldots, W, (J) \tag{7.15}$$

$$\widetilde{n}_{wi}, \text{integer} \geq 0, \ w(i) = 1, 2, \ldots, W, (I) \tag{7.16}$$

Eqs. (7.17−7.19) are the variables that represent the Boolean consideration to determine the existence of any digester, farm operation, or the parking for the standby truck.

$$\widetilde{z}_j \in \{0, 1\}, \quad j = 1, \ldots, J \tag{7.17}$$

$$\widetilde{z}_k \in \{0, 1\}, \quad k = 1, \ldots, K \tag{7.18}$$

$$\widetilde{z}_w \in \{0, 1\}, \ w = 1, \ldots, W \tag{7.19}$$

Eqs. (7.20−7.22) are inequality conditions, which confirm that the digester and the farm area operation exit. This is possible when positive flow to the digester station and the farms is achieved.

$$a\widetilde{X}_{ij} \leq \widetilde{Q}_j \widetilde{z}_j, \quad i(j) = 1, 2, \ldots, I, (J) \tag{7.20}$$

$$a\widetilde{X}_{ik} \leq \widetilde{Q}_k \widetilde{z}_k, \quad i(k) = 1, 2, \ldots, I, (K) \tag{7.21}$$

$$a\widetilde{Y}_{jk} \leq \widetilde{Q}_k \widetilde{z}_k, \quad i(k) = 1, 2, \ldots, I, (K) \tag{7.22}$$

The equations that follow in the following indicate (1) quantity of agrowaste transported to the biodigester Eq. (7.23), (2) quantity of all categories of waste resources transported to the farm Eq. (7.24), (3) quantity of the biodegradable waste released in the farm/landfill area, daily Eq. (7.25), (4) the overall amount of daily agrowaste collected Eq. (7.26). Also, the total number of operational trucks used daily for the waste management system Eq. (7.27) and while on standby Eq, (7.28) per day.

$$\widetilde{w}_j = \sum_i \alpha \widetilde{X}_{ij}, \qquad j = 1, \ldots, J \tag{7.23}$$

$$\widetilde{w}_j = \sum_i \alpha \widetilde{X}_{jk}, \qquad k = 1, \ldots, K \tag{7.24}$$

$$\widetilde{t}_k = \widetilde{w}_k + \sum_j \alpha \widetilde{Y}_{jk}, \quad k = 1, \ldots, K \tag{7.25}$$

$$W = \sum_j \widetilde{w}_j + \sum_k \widetilde{w}_k \tag{7.26}$$

$$T_n + T_a = \sum_{ij} \widetilde{x}_{ij} + \sum_{ik} \widetilde{x}_{ik} + \sum_{ik} \widetilde{y}_{jk} \tag{7.27}$$

$$T_a = \sum_{wj} \widetilde{n}_{wj} + \sum_{ri} \widetilde{n}_{wi} \tag{7.28}$$

Mixed-integer programming

Based on the present research, the same variables as in the linear integer programming are adopted for the mixed-integer programming. However, new sets of continues variables that pertain to the volume of bioresources being transported/time/truck within any of the access nodes are introduced for differentiative purpose, defined by $\widetilde{u}_{ij}, \widetilde{u}_{ik}$, and $\widetilde{v}_{jk}$. Hence, they are used in the proposition to complete the mixed-integer programming. One condition that satisfies the assumption that a truckload is the actually load (full) at any given time is used to address the linear integer approach, whereas the application of the continuous variables is used to determine the resolve based on the true quantity of the wastes to be transported. In essence, the nonnegative inequality condition that confirms that the digester and the farm area operations actually exit is reported in constraints 7.29–7.31 and describes the total wastes from each point, which must not be higher than the capacity of the truck to deliver full load.

$$\widetilde{u}_{ij} \leq \widetilde{Q}_j \widetilde{z}_j, \quad i(j) = 1, 2, \ldots, I, (J) \tag{7.29}$$

$$\widetilde{u}_{ik} \leq \widetilde{Q}_k \widetilde{z}_k, \quad i(k) = 1, 2, \ldots, I, (K) \tag{7.30}$$

$$\widetilde{v}_{ik} \leq \widetilde{Q}_k \widetilde{z}_k, \quad j(k) = 1, 2, \ldots, I, (K) \tag{7.31}$$

The daily amount of agrowastes that reach the biodigester station can be estimated using Eq. (7.32). The following equations indicate (1) quantity of all categories of waste resources transported to the farm Eq. (7.33), (2) quantity of the biodegradable waste released in the farm/landfill area, daily Eq. (7.34), (3) the overall amount of daily agrowaste collected Eq. (7.35), Also, the total number of operational trucks used daily for the waste management system Eq. (7.36) and while on standby Eq. (7.37) per day.

$$\widetilde{w}_j = \sum_i \widetilde{u}_{ij}, \qquad j = 1, \ldots, J \tag{7.32}$$

$$\widetilde{w}_k = \sum_i \widetilde{u}_{ik}, \qquad k = 1, \ldots, K \tag{7.33}$$

$$\widetilde{t}_k = \widetilde{w}_k + \sum_j \widetilde{v}_{jk}, \quad k = 1, \ldots, K \tag{7.34}$$

$$W = \sum_j \widetilde{w}_j + \sum_k \widetilde{w}_k \tag{7.35}$$

$$T_n + T_a = \sum_{ij} \widetilde{x}_{ij} + \sum_{ik} \widetilde{x}_{ik} + \sum_{ik} \widetilde{y}_{jk} \tag{7.36}$$

$$T_a = \sum_{wj} \widetilde{n}_{wj} + \sum_{ri} \widetilde{n}_{wi} \tag{7.37}$$

To determine the factor for the total industrial agrowaste disposal that ought to be daily resituated to the landfills, the investment and operational cost and the base revenue from the energy station were considered. The information received can be imputed as the essential data required to ascertain the minimum cost objective for the waste management system. The latter is possible through the use of any dedicated optimization tool with the capability to analyze linear, nonlinear, and integer programming-associated data (Lee, Yeung, Xiong, & Chung, 2016). Nascent research in the area discussed now considers the chance constrain of early times (Lindo Systems Inc., 2018), which allows for a far-reaching solution via a stochastic analytic process to align newer perspective and establish better economic and environmental conditions for effective, clean energy production operations and waste to wealth realization.

7.4 Conclusion and perspectives

The need to have a well-designed process and approaches that will ensure social, economic, and environmental validation, in line with suitable and cost-effective technologies or strategies in the food and agroindustrial sector, must be given stern consideration. The technologies that support the WTE process have been categorized into two: biological and thermal techniques, which include anaerobic digestion and fermentation and incineration, pyrolysis, and gasification. The exemplification of the integer linear and mixed-integer model has been demonstrated to complement the decision-making process for cost-effective waste management operation. The model covers infrastructural requirements and consideration for the decision on operational capacity about the design (biodigester), the flow of the wastes, collection hot spots, anaerobic digestion station, farm (soil undergoing repair), and the standby truck parking (space). The models are recommended to be applied to promote sustainable waste management operations for contribution to economic, social, and environmental benefits. The collection of accurate data and the application of logical assumptions are pivotal for the success and implementation of the models.

References

Abas, N., Kalair, A., & Khan, N. (2015). Review of fossil fuels and future energy technologies. *Futures*. https://doi.org/10.1016/j.futures.2015.03.003.

Abdel-Shafy, H. I., & Mansour, M. S. M. (2018). Solid waste issue: Sources, composition, disposal, recycling, and valorization. *Egyptian Journal of Petroleum*. https://doi.org/10.1016/j.ejpe.2018.07.003.

Al-Salem, S. M., Lettieri, P., & Baeyens, J. (2009). Recycling and recovery routes of plastic solid waste (PSW): A review. *Waste Management*. https://doi.org/10.1016/j.wasman.2009.06.004.

Babayemi, J. O., Ogundiran, M. B., & Osibanjo, O. (2017). Current levels and management of solid wastes in Nigeria. *Environmental Quality Management*. https://doi.org/10.1002/tqem.21498.

Badran, M. F., & El-Haggar, S. M. (2006). Optimization of municipal solid waste management in Port said – Egypt. *Waste Management*. https://doi.org/10.1016/j.wasman.2005.05.005.

Bhowmik, C., Bhowmik, S., Ray, A., & Pandey, K. M. (2017). Optimal green energy planning for sustainable development: A review. *Renewable and Sustainable Energy Reviews, 71*, 796–813. https://doi.org/10.1016/j.rser.2016.12.105.

Brunner, P. H., & Rechberger, H. (2015). Waste to energy – key element for sustainable waste management. *Waste Management*. https://doi.org/10.1016/j.wasman.2014.02.003.

Dai, C., Li, Y. P., & Huang, G. H. (2011). A two-stage support-vector-regression optimization model for municipal solid waste management – a case study of Beijing, China. *Journal of Environmental Management*. https://doi.org/10.1016/j.jenvman.2011.06.038.

Das, S., & Bhattacharyya, B. K. (2015). Optimization of municipal solid waste collection and transportation routes. *Waste Management*. https://doi.org/10.1016/j.wasman.2015.06.033.

Davis, S. C., Kauneckis, D., Kruse, N. A., Miller, K. E., Zimmer, M., & Dabelko, G. D. (2016). Closing the loop: Integrative systems management of waste in food, energy, and water systems. *Journal of Environmental Studies and Sciences*. https://doi.org/10.1007/s13412-016-0370-0.

Demirbas, A. (2011). Waste management, waste resource facilities and waste conversion processes. *Energy Conversion and Management*. https://doi.org/10.1016/j.enconman.2010.09.025.

Di Bella, G., Di Trapani, D., Mannina, G., & Viviani, G. (2012). Modeling of perched leachate zone formation in municipal solid waste landfills. *Waste Management*. https://doi.org/10.1016/j.wasman.2011.10.025.

Edalatpour, M. A., Al-e-hashem, S. M. J. M., Karimi, B., & Bahli, B. (2018). Investigation on a novel sustainable model for waste management in megacities: A case study in tehran municipality. *Sustainable Cities and Society*. https://doi.org/10.1016/j.scs.2017.09.019.

Ezejiofor, T. (2014). Waste to wealth- value recovery from agro-food processing wastes using biotechnology: A review. *British Biotechnology Journal*. https://doi.org/10.9734/bbj/2014/7017.

Fiorucci, P., Minciardi, R., Robba, M., & Sacile, R. (2003). Solid waste management in urban areas: Development and application of a decision support system. *Resources, Conservation and Recycling*. https://doi.org/10.1016/S0921-3449(02)00076-9.

Fotidis, I. A., Karakashev, D., Kotsopoulos, T. A., Martzopoulos, G. G., & Angelidaki, I. (2013). Effect of ammonium and acetate on methanogenic pathway and methanogenic community composition. *FEMS Microbiology Ecology*. https://doi.org/10.1111/j.1574-6941.2012.01456.x.

Ghasemi, M. K., & Yusuff, R. B. M. (2016). Advantages and disadvantages of healthcare waste treatment and disposal alternatives: Malaysian scenario. *Polish Journal of Environmental Studies*. https://doi.org/10.15244/pjoes/59322.

Ghiani, G., Laganà, D., Manni, E., Musmanno, R., & Vigo, D. (2014). Operations research in solid waste management: A survey of strategic and tactical issues. *Computers and Operations Research*. https://doi.org/10.1016/j.cor.2013.10.006.

Gupta, R. B., & Demirbas, A. (2010). *Gasoline, diesel and ethanol biofuels from grasses and plants*. https://doi.org/10.1017/CBO9780511779152.

Heidenreich, S., & Foscolo, P. U. (2015). New concepts in biomass gasification. *Progress in Energy and Combustion Science*. https://doi.org/10.1016/j.pecs.2014.06.002.

Innovation Strategies in the Food Industry. (2016). *In innovation strategies in the food industry*. https://doi.org/10.1016/c2015-0-00303-3.

Khan, M. D., Khan, N., Sultana, S., Joshi, R., Ahmed, S., Yu, E., … Khan, M. Z. (2017). Bioelectrochemical conversion of waste to energy using microbial fuel cell technology. *Process Biochemistry*. https://doi.org/10.1016/j.procbio.2017.04.001.

Kirtania, K., & Bhattacharya, S. (2012). Application of the distributed activation energy model to the kinetic study of pyrolysis of the fresh water algae *Chlorococcum humicola*. *Bioresource Technology*. https://doi.org/10.1016/j.biortech.2011.12.094.

Kirtania, K., & Bhattacharya, S. (2015). Coupling of a distributed activation energy model with particle simulation for entrained flow pyrolysis of biomass. *Fuel Processing Technology*. https://doi.org/10.1016/j.fuproc.2015.04.014.

Lawal, A. Q. T., Ninsiima, E., Odebiyib, O. S., Hassan, A. S., Oyagbola, I. A., Onu, P., & A, D. (2019). Effect of unburnt rice husk on the properties of concrete. *Procedia Manufacturing, 35*, 635–640. https://doi.org/10.1016/j.promfg.2019.06.006.

Lee, C. K. M., Yeung, C. L., Xiong, Z. R., & Chung, S. H. (2016). A mathematical model for municipal solid waste management – a case study in Hong Kong. *Waste Management*. https://doi.org/10.1016/j.wasman.2016.06.017.

Lega, M., D'Antonio, L., & Napoli, R. M. A. (2010). Cultural Heritage and Waste Heritage: Advanced techniques to preserve cultural heritage, exploring just in time the ruins produced by disasters and natural calamities. *WIT Transactions on Ecology and the Environment*. https://doi.org/10.2495/WM100121.

Liebetrau, J., Sträuber, H., Kretzschmar, J., Denysenko, V., & Nelles, M. (2019). Anaerobic digestion. In *Advances in biochemical engineering/biotechnology*. https://doi.org/10.1007/10_2016_67.

Lindo Systems Inc.. (2018). *The modeling language and optimizer*. Lindo Systems Inc.

Lyu, Z., Shao, N., Akinyemi, T., & Whitman, W. B. (2018). Methanogenesis. *Current Biology*. https://doi.org/10.1016/j.cub.2018.05.021.

Mavrotas, G., Skoulaxinou, S., Gakis, N., Katsouros, V., & Georgopoulou, E. (2013). A multi-objective programming model for assessment the GHG emissions in MSW management. *Waste Management*. https://doi.org/10.1016/j.wasman.2013.04.012.

Meena, M. D., Yadav, R. K., Narjary, B., Yadav, G., Jat, H. S., Sheoran, P., … Moharana, P. C. (2019). Municipal solid waste (MSW): Strategies to improve salt affected soil sustainability: A review. *Waste Management*. https://doi.org/10.1016/j.wasman.2018.11.020.

Miandad, R., Barakat, M. A., Aburiazaiza, A. S., Rehan, M., & Nizami, A. S. (2016). Catalytic pyrolysis of plastic waste: A review. *Process Safety and Environmental Protection*. https://doi.org/10.1016/j.psep.2016.06.022.

Onu, P., Abolarin, M. S., & Anafi, F. O. (2017). Assessment of effect of rice husk ash on burnt properties of badeggi clay. *International Journal of Advanced Research, 5*(5), 240–247. https://doi.org/10.21474/IJAR01/4103.

Onu, P., & Mbohwa, C. (2018a). Future energy systems and sustainable emission control: Africa in perspective. *Proceedings of the International Conference on Industrial Engineering and Operations Management*.

Onu, P., & Mbohwa, C. (2018b). Sustainable oil exploitation versus renewable energy Initiatives: A review of the case of Uganda. *Proceedings of the International Conference on Industrial Engineering and Operations Management*, 1008–1015 (Washington DC).

Onu, P., & Mbohwa, C. (2019a). Sustainable production: New thinking for SMEs. *Journal of Physics: Conference Series, 1378*(2). https://doi.org/10.1088/1742-6596/1378/2/022072.

Onu, P., & Mbohwa, C. (2019b). Sustainable supply chain management: Impact of practice on manufacturing and industry development. *Journal of Physics: Conference Series, 1378*(2). https://doi.org/10.1088/1742-6596/1378/2/022073.

Onu, P., & Mbohwa, C. (2019c). *Advances in solar photovoltaic grid parity.* https://doi.org/10.1109/IRSEC48032.2019.9078175.

Onu, P., & Mbohwa, C. (2019d). Renewable energy technologies in brief. *International Journal of Scientific & Technology Research, 8*(10), 1283–1289.

Onu, P., & Mbohwa, C. (2019e). Industrial energy conservation initiative and prospect for sustainable manufacturing. *Procedia Manufacturing, 35*, 546–551. https://doi.org/10.1016/j.promfg.2019.05.077.

Onu, P., & Mbohwa, C. (2019f). *New future for sustainability and industrial development : Success in blockchain, internet of production, and cloud computing technology* (pp. 1–12).

Peterson, A. A., Vogel, F., Lachance, R. P., Fröling, M., Antal, M. J., & Tester, J. W. (2008). Thermochemical biofuel production in hydrothermal media: A review of sub- and supercritical water technologies. *Energy and Environmental Science*. https://doi.org/10.1039/b810100k.

Pham, T. P. T., Kaushik, R., Parshetti, G. K., Mahmood, R., & Balasubramanian, R. (2015). Food waste-to-energy conversion technologies: Current status and future directions. *Waste Management*. https://doi.org/10.1016/j.wasman.2014.12.004.

Reichenbach, J. (2008). Status and prospects of pay-as-you-throw in Europe – a review of pilot research and implementation studies. *Waste Management*. https://doi.org/10.1016/j.wasman.2008.07.008.

Ross, E. (2018). *How tech is saving the environment.*

Santoro, C., Arbizzani, C., Erable, B., & Ieropoulos, I. (2017). Microbial fuel cells: From fundamentals to applications. A review. *Journal of Power Sources*. https://doi.org/10.1016/j.jpowsour.2017.03.109.

Schlüter, A., Bekel, T., Diaz, N. N., Dondrup, M., Eichenlaub, R., Gartemann, K. H., … Goesmann, A. (2008). The metagenome of a biogas-producing microbial community of a production-scale biogas plant fermenter analysed by the 454-pyrosequencing technology. *Journal of Biotechnology*. https://doi.org/10.1016/j.jbiotec.2008.05.008.

Sharma, A., Pareek, V., & Zhang, D. (2015). Biomass pyrolysis – a review of modelling, process parameters and catalytic studies. *Renewable and Sustainable Energy Reviews*. https://doi.org/10.1016/j.rser.2015.04.193.

Steinbusch, K. J. J., Arvaniti, E., Hamelers, H. V. M., & Buisman, C. J. N. (2009). Selective inhibition of methanogenesis to enhance ethanol and n-butyrate production through acetate reduction in mixed culture fermentation. *Bioresource Technology*. https://doi.org/10.1016/j.biortech.2009.01.049.

Tambone, F., Genevini, P., D'Imporzano, G., & Adani, F. (2009). Assessing amendment properties of digestate by studying the organic matter composition and the degree of biological stability during the anaerobic digestion of the organic fraction of MSW. *Bioresource Technology*. https://doi.org/10.1016/j.biortech.2009.02.012.

Tan, S., Hashim, H., Lee, C., Taib, M. R., & Yan, J. (2014). Economical and environmental impact of waste-T o-energy (WTE) alternatives for waste incineration, landfill and anaerobic digestion. *Energy Procedia*. https://doi.org/10.1016/j.egypro.2014.11.947.

Umeda, K., Nakamura, S., Lu, D., & Yoshikawa, K. (2019). Biomass gasification employing low-temperature carbonization pretreatment for tar reduction. *Biomass and Bioenergy.* https://doi.org/10.1016/j.biombioe.2019.05.002.

Vaish, B., Sharma, B., Srivastava, V., Singh, P., Ibrahim, M. H., & Singh, R. P. (2019). Energy recovery potential and environmental impact of gasification for municipal solid waste. *Biofuels*. https://doi.org/10.1080/17597269.2017.1368061.

Walter Borges de Oliveira, S. V., Leoneti, A. B., Magrini Caldo, G. M., & Borges de Oliveira, M. M. (2011). Generation of bioenergy and biofertilizer on a sustainable rural property. *Biomass and Bioenergy.* https://doi.org/10.1016/j.biombioe.2011.02.048.

Welfle, A., Gilbert, P., Thornley, P., & Stephenson, A. (2017a). Generating low-carbon heat from biomass: Life cycle assessment of bioenergy scenarios. *Journal of Cleaner Production*. https://doi.org/10.1016/j.jclepro.2017.02.035.

Welfle, A., Gilbert, P., Thornley, P., & Stephenson, A. (2017b). Generating low-carbon heat from biomass. *Journal of Cleaner Production.* https://doi.org/10.1016/j.jclepro.2017.02.035.

Yau, Y. (2010). Domestic waste recycling, collective action and economic incentive: The case in Hong Kong. *Waste Management.* https://doi.org/10.1016/j.wasman.2010.06.009.

Yusuf, A. A., Peter, O., Hassanb, A. S.,A.,L. T., Oyagbola, I. A., Mustafad, M. M., & Yusuf, D. A. (2019). Municipality solid waste management system for Mukono district. *Procedia Manufacturing, 35*, 613—622. https://doi.org/10.1016/j.promfg.2019.06.003.

Zhang, D., Luo, W., Li, Y., Wang, G., & Li, G. (2018). Performance of co-composting sewage sludge and organic fraction of municipal solid waste at different proportions. *Bioresource Technology.* https://doi.org/10.1016/j.biortech.2017.08.136.

Zong, W., Yu, R., Zhang, P., Fan, M., & Zhou, Z. (2009). Efficient hydrogen gas production from cassava and food waste by a two-step process of dark fermentation and photo-fermentation. *Biomass and Bioenergy.* https://doi.org/10.1016/j.biombioe.2009.06.008.

CHAPTER

Nascent technologies in resources conservation and sustainable agricultural development

8

8.1 Introduction

Environmental degradation challenges have become increasingly obvious and threatened the natural climatic balance. This has drawn attention to ecological security and sustainable intervention, in which ensuring the preservation of the natural habitat and living organism is vital. Several nations across the globe are sorting means to develop their economic strength, to boost their GDP through agricultural expansion while leveraging its operations on social and economic benefits. As such, they provide quality food available for the populist in a sound and healthy environment, thus protecting the ecosystem (Pan, Ai, Li, Pan, & Yan, 2019; Song, Fisher, & Kwoh, 2019; Wang & Song, 2014). In other developing countries, and Africa in perspective to diversify their economy, the prioritization of technoinnovation advancement, with the focus to improve resources conservation and agricultural development initiatives, should be uncontested. More so, it implements a blueprint to deploy the necessary strategies, linked to ecoefficient innovation, aimed at effective agriculture diversity.

The aforementioned are potentials for sustainable development vis-à-vis economic and environmental protection while eliminating material wastages with savings on energy and reduction in emissions (Lawal et al., 2019; Yusuf et al., 2019; Peter & Mbohwa, 2019). The definition of green innovation has been provided according to notable researchers, presented in Table 8.1. The concept and application of the green innovation techniques are meant to tackle the lack of efficient practices and irrational use of the abundant resource, thus considering the effectiveness and reduction of the different dimensions of resource conservation and environmental appraisal.

Agricultural waste diversity: conception and sustainable approach to urbanization

Accordingly, sustainability with regard to the triple bottom line consideration (social environment and economic) focuses on the 3P: people, planet, and profit,

Agricultural Waste Diversity and Sustainability Issues. https://doi.org/10.1016/B978-0-323-85402-3.00011-5

Table 8.1 Definition of green innovation.

	Definition of green innovation	Reference
1	"New products and processes which provide customer and business value but significantly decrease environmental impacts."	Fussler (1996)
2	"The production, assimilation or exploitation of a product, production process, service or management or business method that is novel to the organization and which results, throughout its life cycle, in a reduction of environmental risk, pollution and other negative impacts of resources use (including energy use) compared to relevant alternatives."	Kemp and Pearson (2007)
3	"As innovations that consist of new or modified processes, practices, systems, and products which benefit the environment and so contribute to environmental sustainability."	Oltra and Saint Jean (2009)
4	"As hardware or software innovation that is related to green products or processes, including the innovation in technologies that are involved in energy-saving, pollution-prevention, waste recycling, green product designs, or corporate environmental management."	Chang (2011)

which was first conceived in 1987 (Elkington, 2004, Elkington, 2013) and ensued by the deliberation and call for sustainable development disposition. Hence, the Brundtland (1987) proposition: "sustainable development is development that meets the needs of the present without compromising the ability of future generations to meet their own needs." However, the mandate was passed on as a collective responsibility where enterprises design their own sustainable matrix to determine what works for them and according to their scope of operation. As such, a recommendation of the Brundtland (1987) deliberation is to be extended to institutions and researched for future sustainable development, with a focus on societal, social, economic, and environmental consideration while promoting awareness about conservation of the ecosystem. The application of these principle guides activities which pertain to the management of wastes in all its' forms, while preserving the naturally occurring resources (Moldan, Janoušková, & Hák, 2012; Morelli, 2011).

The appraisal of sustainable initiatives within the agroindustrial sector is gradually coming to light and has received a considerable contribution. This is as a result of the urgent need to diversify the global practice of fossil fuel exploration and conservation of natural resources. Hence, conscience research and attention are directed toward innovation and sustainable development practices, to ascribe importance to environmental perseverance and social satisfaction (Albort-Morant, Leal-Millán, & Cepeda-Carrión, 2016; Boons & Lüdeke-Freund, 2013; Matos & Silvestre, 2013).

New systems, operational protocols, and models and training of experts to drive the technological and service protocols, based on the sustainability dimension and interest of any designated enterprise are needed (Onu & Mbohwa, 2018a). The ecoinnovative approach for sustainability ought to address material/resource cost reduction, savings on energy, and the use of alternative sources over long periods (Mamede & Gomes, 2014). So far, the emphasis has been on regulatory considerations and incentivizations used as drivers for sustainable development (Hojnik & Ruzzier, 2016). The focus is on agrarian activities and the conservation of green resources for the promotion of agrowaste management linked to the triple bottom line indicators.

Furthermore, the premise of natural resource scarcity in some countries around the world, compared with others, draws concern, in that while tantamount efforts are being made in one part of the world to promote sustainable development with regard to the triple bottom line standards, the rest are lagging behind (Iraldo, Testa, & Frey, 2009). This analogy is described from the perspective illustrated in Fig. 8.1. Nations classified to be developing countries must expedite the call for sustainability advancement and charge for research, development, deployment, and demonstration

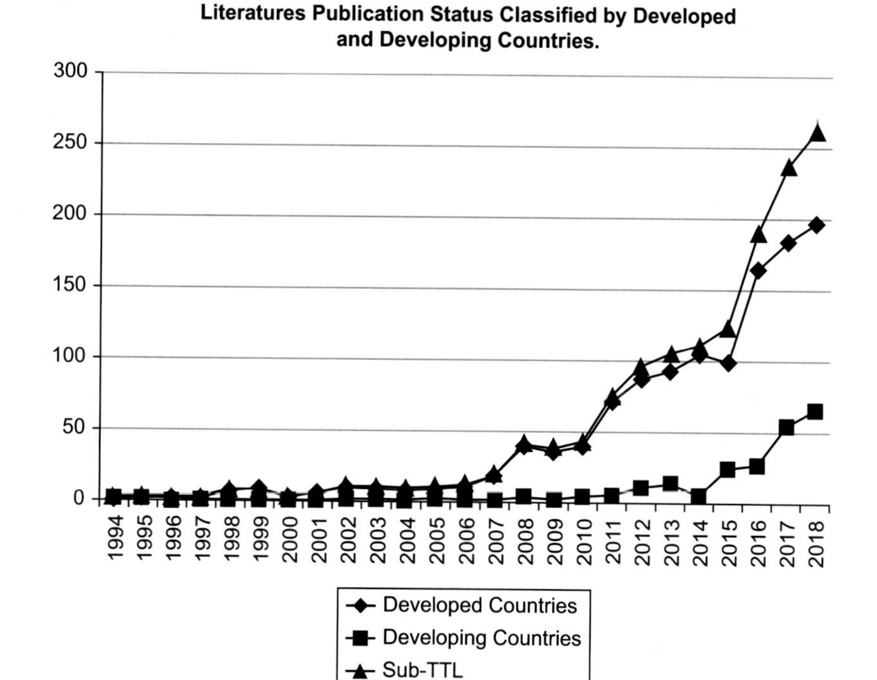

FIGURE 8.1

Research/scientific input on ecoinnovation (1994–2018) by developed and developing countries. *Sub-TTL*, subtotal.

Adapted from Green and low-carbon technology innovations. Innovation Strategies in Emerging Science 2020 209–253; Lai, X., & Shi, Q. (2020). Green and low-carbon technology innovations.

to contribute to the global push, hence addressing the eminent natural resource depletion dilemma to meet the needs of future social, economic, and environmental development (Rimos, Hoadley, & Brennan, 2014).

Concept of data management in sustainable agricultural development

The aspect of technological advancement that propagates data management consideration has sprung up new opportunities, which indirectly challenges the concept of ecoefficiency as well. While it is essential to support the green technology trajectory to align with organizational activities and meet corporate business goals, the idea of nascent technoinnovation promulgation that will guide the implementation process is important. Data management is an emerging path in informatics services that offer new knowledge, values, and capabilities through the integration of seemingly unclear and fragmented information related to environmental conditions to apply it to a solution-oriented approach. Data management promotes organizational effectiveness and efficiency of environmental governance where openness in sharing data helps to facilitate synergetic interactions and deliver corporate goals (Peter & Mbohwa, 2018b). These actions can be directed toward the improvement of agricultural development and resource waste conservation.

The application of data analytics in agronomy has helped to improve the effectiveness of activities relating to ecoinnovation and promote modern operations of sustainable development indicators (Can & Alatas, 2017). Access to the technologies, and knowledge requirement to operate them, or the cost associated with the acquisition of these technologies is a vital issue to be addressed (Onu & Mbohwa, 2019c, 2019b, 2018a). Already, large pile of data and digital material exist with reference to global agricultural activities that can bolster operational effectiveness, adapted from records and monitoring, or as in recent times the use of global positioning systems (GPS) (Huang, Chen, Yu, Huang, & Gu, 2018; Morota, Ventura, Silva, Koyama, & Fernando, 2018; Tien, 2013). The information recouped from the aforementioned protocol can transform how agricultural development practitioners may apply the strategy to increase productivity.

Large-scale data management has the potential to promote ecoinnovation and decision regarding the level of supports required to guarantee sustainable agricultural activities that will lead diversity in the sector (Bibri, 2018; Cheng, Yang, & Sheu, 2014; De Gennaro, Paffumi, & Martini, 2016; OECD, 2010; Smol, Kulczycka, & Avdiushchenko, 2017). Data analytics is no doubt being appreciated in small, medium, and big enterprises, as both developing and underdeveloped countries are learning about its application and strategies through which the knowledge can be exploited to transition agricultural development (Hampton et al., 2013). Hence, in the near future, it offers solutions to shortcomings on effective resources management and information regarding labor consideration, materials pricing, resource endowment, weather, land conditions, air pollution metrics, and technological awareness, among other considerations. In essence, essential data can now be accessed on different database within several platforms and accessible on the Internet (United States Environmental Protection Agency, 2014; World Bank, 2016). The efficient utilization of data management protocol is important for solving the widespread global challenges, due to environmental pollution and agro-based resources conservation.

8.2 Green innovation strategy and implementation success: the perspectives of agricultural resources development

The green innovation concept is directly linked to environmental solutions and leads to the realization of corporate benefits as it pertains to sustainable development (Weng, Chen, & Chen, 2015). Literature suggests a positive correlation between operational performance and green innovation implementation (Hassan et al., 2016; Yim, Fung, & Lau, 2010; Yu, Chavez, Feng, & Wiengarten, 2014); its tendencies can contribute to environmental management and pollution preventive achievement used to measure performance (Chang, 2011; Kammerer, 2009). The process of applying green innovation and the services involved impact the environment positively with a noticeable improvement in social and economic welfares, by virtue of the reduction in wastes and cost of the waste management operations (Zeng & Stratton, 2017). Hence, the urgency to implement green innovation in the agricultural sector requires speedy action to transform productivity, as this has been applied in the manufacturing industrial spaces for timeous and cost reduction operations.

This chapter refers to green innovation as a wordstream, which has been used interchangeably used as ecotechnology and green technology (Schiederig, Tietze, & Herstatt, 2012), thus depicting in a broader sense nascent activities, which may either be technological or nontechnological; however, its implementation can foster operational effectiveness, business competitiveness, and resource management, in line with environmental protection and investors, and consumers interest. More so, where the activities are sustainably inclined and environmentally friendly, the phenonium is regarded as a green innovation process (Fei, Wang, Yang, Chen, & Zhi, 2016; Kunapatarawong & Martínez-Ros, 2016).

Recently, researchers (Lorente & Álvarez-Herranz, 2016; Wang, Bao, Wen, & Tan, 2016) demonstrated the use of "environmental Kuznets curve," to ascertain a positive correlation between economic growth potential and ecological protection. The fuzz about green innovation on energy conservation continues to elude researchers, and there is yet to be any permanent dimension or understanding regarding technology or its deployment (Abdu, Peter, & Gupta, 2016; Onu & Mbohwa, 2018b). Relating to technological adoption in agrowaste management, its essence is to track, monitor, prevent, and control pollution during the operation to ensure adequate, social, and environmental risk-aversive and cost-effective process. Although the opportunities for green technology adoption hold potentials for energy savings and adequate resources management, current traditional methods cannot be quickly phased even though they are nongreen technologies and offer a less sustainable solution. The onus lies on researchers to find pathways to develop or optimize present-day technologies and to search the most economical means to replace them with a fair level of guarantee on return on investment.

There is a thin line between sustainable development and resource conservation, as the two are on par. However, most countries have become great achievers in the socioeconomic aspect, performing poorly on environmental and societal achievement. A typical example is comparing China's excellence and remarkable development in terms of social amenities and economic transformation in the past decades, regardless of their population. China may not be performing as such in terms of environmental expectations, caused by limited natural resource reserve; the communist country continues to increase their renewable energy interactions to promote the overall climate situation of the country. Africa, on the other hand, is at the brink of calamity, at the current rate of natural resources exploitation and dependence on fossils, which undermines their sustainability effort. The need to consolidate and tackle the resources depletion dilemma must, therefore, sort to embrace sustainable natural resource management and green technology advancement (Rimos et al., 2014).

The agrowaste management process provides a sustainable route to resources management, which also considers future aspects and benefits to the generation yet to come (Mustapha, Manan, & Wan Alwi, 2017). While there is a dare need to expand sustainable agricultural development, the practices ought to lead efficient maximization of resources and utilization, leaving no wastes (from wastes to energy and production of biofuels and other materials) through a standard operation process that recognized values (Wagner, 2015). The improvement of natural resource management, through the development, utilization, and perseverance, is tantamount toward the triple bottom line achievement and supports sustainable development strategies (Peter & Mbohwa, 2018a). The solutions to issues surrounding natural resources management, the economics of environmental preservation, and ecoinnovation will become the building blocks for accessing near-future sustainability achievement (countries and regions) (Akiyode, Tumushabe, Hadijjah, & Peter, 2017). It is pertinent to emphasize the gain due to the possibilities and potential of generating sustainable electricity from the natural resource wastes (agriculture food residues within cities and farms), as a waste management endeavor (Chalvatzis, Malekpoor, Mishra, Lettice, & Choudhary, 2019; Deng & Gibson, 2019).

Also, the role of researchers toward publications, and raising awareness of the factors that influence technological innovation and energy efficiency, thus contributing to ecoefficiency implementation and green supply chain operation, is essential for environmental economics to promote sustainability (Kumar & Chandrakar, 2012; Mustapha et al., 2017; Ruan et al., 2019; Shen, Choi, & Chan, 2019). A recent contribution to green innovation management suggests the promulgation of e-waste recycling (Luhar & Luhar, 2019). From the agricultural development context, data analysis should be made available to guide operations and stakeholders' decisions on farming activities (Garlapati, 2016; Song, Cao, & Wang, 2019; Tansel, 2017).

Categories of innovation for agricultural sustainable solution

The nature of innovativeness and intervention required to advance sustainability and proffer solution about environmental challenges and issues linked to agrowastes seeks support from all the stakeholders affected (Canonico, Consiglio, Iacono, Mercurio, & Berni, 2014; Mor, Singh, Bhardwaj, & Singh, 2015; Sousa-Zomer & Cauchick Miguel, 2018). Researchers are relentless about drivers for the adoption of ecoinnovation and what may guide the decision of different organizations to make a move on what innovative approach to be taken (Bossle, Dutra De Barcellos, Vieira, & Sauvée, 2016; Hojnik & Ruzzier, 2016). More so, the dichotomy between sustainable, environmental, and ecoinnovation has received progressive research (Bellotti & Rochecouste, 2014; Cuerva, Triguero-Cano, & Córcoles, 2014; Dogliotti et al., 2014; Viaggi, 2015). The innovation conception comprises the technology/technical, service/process, and the business perspective for performance improvement and implementation (Junior, Oliveira, & Yanaze, 2019; Lai & Shi, 2020; Onu & Mbohwa, 2019a).

Technology innovation from the perspective of agricultural sustainable solutions adopts nascent techniques/strategies for green products creation and development of renewable energy systems (Onu & Mbohwa, 2019b). Service innovation involves the renewal and customization of processes for effective service deliveries. This has been recommended as the timeliest intervention for transforming the agricultural sector (Koskela-Huotari, Edvardsson, Jonas, Sörhammar, & Witell, 2016). According to Yang, Evans, Vladimirova, and Rana (2017), the business model innovation does not necessarily discover a new product or service; instead, it uses unique ways to create and promote existing products, or services and new ways to capture values from them. A model for determining technical and ecoefficiency is presented in the subsequent section as part of the green innovation strategy for agricultural development.

Stochastic frontier approach: determination of technical and ecoefficiency

A demonstration using the stochastic frontier approach to analyze ecoefficiency is presented with the aim to guide sustainability implementation decisions and green innovation. The concept first sufficed in 1977 as the stochastic production frontier model (Aigner, Lovell, & Schmidt, 1977; Meeusen & van Den Broeck, 1977) has been applied to evaluate ecological performance indicators and ecoefficiency over the years (Berre et al., 2015; Greene, 2008; Robaina-Alves, Moutinho, & MacEdo, 2015). The practice of ecoefficiency analysis is meant to access the performance of environmental and ecological factors that satisfy societal expectations (Tamayo-Orbegozo, Vicente-Molina, & Villarreal-Larrinaga, 2017). The perception is more toward the capability of developing economies to embrace models that can transform their farming experience while improving the overall resource management and agricultural development process. Stochastic frontier has been demonstrated in the present section to enable the analysis of two perspectives: (1) technical efficiency and (2) ecoefficiency. Multiinput, output production function, including the ecological and land variables as an input function, in a possible farming operations assessment, is defined. Assuming period $t = 1, \ldots, T$, $X^t \in R_+{}^N$ *and* $Y^t \in R_+{}^N$ represent input and output vectors, then:

$$S^t = \{(X^t, Y^t): X^t \text{can produce } Y^t\} \tag{8.1}$$

Also, considering the distance function methodology, according to authors (Song, An, Zhang, Wang, & Wu, 2012), the input distance function is expressed as in Eq. (8.2), so that the input vector X^t is adjusted over the output Y^t to test for technological feasibility. But, only if $D_I{}^t(X^t, Y^t) \leq 1$ holds, or if and only if $(X^t, Y^t) \in S^t$ conditions are met:

$$D_I^t(X^t, Y^t) = \sup\{\lambda: (X^t / \lambda, Y^t) \in S^t\} \tag{8.2}$$

Furthermore, $D_I{}^t(X^t, Y^t) = 1$ if and only if (X^t, Y^t) satisfies the boundary conditions (frontier of technology). Hence, for the *ith,* observations, Eq. (8.3) defines the stochastic frontier model that can be used.

$$D_I^t(X_i^t, Y_i^t, t; \alpha, \beta, \gamma, \varphi, \delta, \phi)\exp(v_i - u_i) = 1 \tag{8.3}$$

The parameters α, β, γ, φ, δ, and ϕ are taken as unknown at this stage but are estimated subsequently.

The expression v_i is introduced to satisfy noise consideration. u_i is the expression for the technical inefficiency model, so that *i.i.d* $v_i{}^t \sim N(0, \sigma_u^2)$ and $u_i{}^t \sim N^+(u_i, \sigma_u^2)$, and is rewritten as Eq. (8.4).

$$u_i = \tau_0 + Z_{ij} * \tau_j \tag{8.4}$$

Z_{ij} represents the vector that describes the technical inefficiency effects, and τ_0 is the constant of the model which is a transformation of Eq. (8.3) to give:

$$\ln\left(D_I^t(X_i^t, Y_i^t, t)\right) = u_i - v_i \tag{8.5}$$

For the normalization of the distance function, in the case of homogeneity characteristics, inputs can be rewritten as:

$$D_I^t(X_i^t / x_n, Y_i^t, t) = D_I^t(X_i^t, Y_i^t, t)/x_n$$

$$-\ln x_n = \ln\left(D_I^t(X_i^t / x_n, Y_i^t, t)\right) - \ln\left(D_I^t(X_i^t, Y_i^t, t)\right) \tag{8.6}$$

Thus, transforms Eqs (8.5) and (8.6) to become an expression that can be used to estimate technical efficiency (Eq. 8.7).

$$-\ln x_n = -\ln\left(D_I^t(X_i^t / x_n, Y_i^t, t)\right) - u_i + v_i \tag{8.7}$$

Technical efficiency is defined as: "The ration of the observation output to the corresponding potential output given production frontier" according to Song et al. (2013), which gives rise to the relationship Eq. (8.8).

$$Y_i = f(X_i, \beta)\exp(v_i - u_i) \tag{8.8}$$

Hence, technical efficiency is rewritten and presented as:

$$TE_i = \frac{f(X_t, \beta)\exp(v_i - u_i)}{f(X_i, \beta)\exp(v_i)} = \exp(-u_i) = \frac{1}{D_I^t(X_i^t, Y_i^t, t)} \tag{8.9}$$

Again, from the work done by Reinhard, Lovell, and Thijssen (1999), ecoefficiency can be calculated by Eq. (8.10).

$$EE_i = \frac{min.feasible\ ecological\ input}{observed\ ecological\ input} \tag{8.10}$$

Hence, let the output distance be Eq. (8.11), in comparison with Eq. (8.2) given earlier.

$$D_O^t(X^t, Y^t) = (\sup\{\lambda : (X^t, \lambda Y^t) \in S^t\})^{-1} \tag{8.11}$$

Eq. (8.11) expresses how the other functions are defined by the expansion of the output and input vectors (X^t, Y^t). The ecological consideration of a case area or cultivated area is a measure of input, denoted as FA_i. Also, the land input described in Eqs. (8.12 and 8.13) can be treated as a recompense between urbanization and land productivity, or the incorporation of land productivity and cropping returns, and urbanization and cropping returns into a net primary production value. As such, ecoefficiency based on agricultural development can be estimated using relations Eqs. (8.14) and (8.15)

$$X_S^t = (cultivatedarea,\ 'FA_i',\ capital,\ labor,\ property\ 1,\ property\ 2, \ldots) \tag{8.12}$$

$$Y_S^t = (land\ productivity) \tag{8.13}$$

$$\begin{cases} -\ln x_n = -\ln\left(D_I^t(X_i^t(FA_i)/x_n, Y_i^t, t)\right) - u_i + v_i \\ -\ln x_n = -\ln\left(D_I^t\left(X_i^t\left(\widehat{FA_i}\right)/x_n, Y_i^t, t\right)\right) - \overline{u_i} + v_i \end{cases}$$

$$EE_i = \frac{'FA_i'}{Y_S^t} \tag{8.14}$$

$$EE_i = \frac{min.feasible\ NPP_i}{NPP_i} \tag{8.15}$$

8.3 Technological considerations, challenges, and policies for green innovation and sustainable development

A recent trend in research has led to the development and use of a dynamic energy systems model to access the impact of the ecoinnovation initiatives on a country's climatic condition and based on the available international policies. Montalvo (2008) published concerning the advent of cleaner technologies having great economic potential and the challenges over time. The authors went ahead to state that the technological intervention will have a considerable impact and require policy changes to register any recognizable difference.

Further, into the early 2010s, Pearson and Foxon (2012), research on the long-term benefits of evolving sustainable initiatives and integration of technologies offers conservative potentials to promote low carbon innovation. In China, an investigation was conducted to determine how several factors, such as investments in energy systems, energy prices, research outputs, and development, impact energy technology patent. The study was carried out in 30 provinces of China and lasted for 5 years and ended in 2013 with the result showing a positive correlation between technological integration and economic growth (Li & Lin, 2016).

In recent years, researchers have tested different models to discern the progress of green economic policies for research and development and its impact on innovative and renewable energy systems (Lindman & Söderholm, 2016). Literature (Alvarez-Herranz, Balsalobre-Lorente, Shahbaz, & Cantos, 2017) shows that contributions have been made regarding carbon emission, energy regulations, and economic growth potential, with an assertion that there exists a positive relationship between the energy innovation process and environment pollution. A list of green technologies that can support low-carbon contributions and serve numerous operations in the agricultural domain, among other sectors, has been identified (Adnan, Nordin, Rahman, Vasant, & Noor, 2017; Söderholm, Bergquist, & Söderholm, 2017). According to Adnan, beyond the conversion of agrowastes to be used as clean fuels, its energy potential is accessible for electrical generation. An investigation into the exploitation of green technological systems and the diffusion tendencies has also been conducted (McGorry, Ratheesh, & O'Donoghue, 2018). Additionally, a timely and comprehensive study has sufficed; Wiseman (2018) carried out the research on the pathways through which a "zero net emissions" can be achieved from a global perspective.

The factors to be considered for renewable energy systems to be diffused, to impact environmental and socioeconomic readiness and perspectives, have been explored (Niero, Olsen, & Laurent, 2018; Yun & Lee, 2015). Al-Ansari (2014) consolidated innovation practices and growth tendencies in the United Arab Emirate, wherein a conceptualized model was used to guide implementation within SMEs to forester innovation and sustainable business growth. The aspect of customer roles and intervention toward adoption or implementation of cleaner technologies and sustainable initiatives has been addressed by (Coad, de Haan, & Woersdorfer, 2009). As such, the promotion of green cars will invariably promote ecoefficiency and policies that support environmental protection, through an exploratory research, which discusses the integration of green innovation and supply chain, emphasizes the need to improve performance. Hence, the concept of resource conservation and zero wastes initiatives must advance to a level where it becomes a household practice for both the supplier and the customer (Wu, 2013). While ecoinnovation appraisal has been seen to usher new solutions to operational challenges and remediate complexities, more is desired administratively to empower those responsible for driving the required organizational practices from the sustainability point of view (Ketata, Sofka, & Grimpe, 2015).

Premise the above, different sustainable initiatives, and strategies that support the application of eco-diversity solutions, and the implementations of innovative approaches, have promoted environmental protection and cleaner energy proposition. In essence, advancement in technology innovation and management science is responsible for the development of resource conservation that will lead to optimal resource utilization for a sustainable ecosystem. Efforts to implement models of technology innovation and policies must become a global priority. This will push research, testing, implementation, and demonstration to overcome the challenges that hinder agricultural development and sustainable development. Dynamic networks and access to ecoinnovation systems are needed to strengthen the connection between sustainable agricultural diversity mechanisms and technology, especially in poor/rural or developing settlements. Information communication systems and knowledge acquisition of the most recent advancement and approach in wastes management must be necessitated.

8.4 Conclusions and perspectives

The purview into nascent technologies that pertain to resource conservation and sustainable agricultural development has been explored in this chapter. While this chapter identifies the need for resources conservation and agricultural development as a means to advance ecoefficiency, it is essential to promote socioeconomic and environmental performance and sustainable values within the agroindustrial sector.

Thus, the authors discuss traditional techniques, while exploring new strategies that are required for sustainable agricultural development, citing models that facilitate the assessment of technical and ecoefficiency and address economic consideration, technological potential, and policies that can promote agrowaste diversity. Emphasis on natural resources management, green growth, environmental governances, and cleaner production initiatives is pivotal to achieve sustainable socioeconomic and environmental development. Hence, the mechanism for implementation must receive significant private–public/government support through partnership and collaboration with stakeholders involved in sustainability or active practitioners in agricultural development. The authors recognize the disparity in the ecoefficiency and green innovation agitation, as they may require varying approaches, skill sets, and models. However, the juxtaposition of four considerations that link green innovation with ecoefficiency to lead sustainability makes inferences to (1) environmental innovation, (2) organizational innovation, (3) operational innovation, and (4) technological innovations. Importantly, the focus on environmentally improved products, processes, and services and the adoption of alternatives/nascent systems with improved performance compared with traditional methods is a timely intervention.

The authors infer that new knowledge development and advancement of social and cultural responsibilities to manage resources both in the township and rural settlement must become the business for all. Resource conservation, wastes management, and the proliferation of technological innovations that respond to socioeconomic improvement and environmental protection cannot prevail without favorable policies. There is a strong connection, therefore, between current evidence of scientific research, to advance sustainable disposition, and future ecosystem development (climatic condition, process optimization, product development, operational performance, and organizational effectiveness, to mention a few). Future directions must consider solutions to critical challenges that pulverize rural engagement and contribution to sustainability. Sub-Saharan African countries, and a list of other developing nations, can extend information technology and prioritize appraisal of e-commerce to rural communities where agricultural development activities are pronounced. This will usher new possibilities and extend innovative capabilities that are desired to simultaneously promote the advancement of agricultural development, environmental protection, and improvement of livelihood.

References

Abdu, Y. A., Peter, O., & Gupta, U. K. (2016). Lean concepts and methods: 3P. *International Journal of Scientific Research in Computer Science, Engineering and Information Technology, 1*(2), 20–24.

Adnan, N., Nordin, S. M., Rahman, I., Vasant, P. M., & Noor, A. (2017). A comprehensive review on theoretical framework-based electric vehicle consumer adoption research. *International Journal of Energy Research*. https://doi.org/10.1002/er.3640.

Aigner, D., Lovell, C. A. K., & Schmidt, P. (1977). Formulation and estimation of stochastic frontier production function models. *Journal of Econometrics*. https://doi.org/10.1016/0304-4076(77)90052-5.

Akiyode, O., Tumushabe, A., Hadijjah, K., & Peter, O. (2017). Climate change, food security and environmental security: A conflict inclination assessment of Karamoja region of Uganda. *International Journal of Scientific World*. https://doi.org/10.14419/ijsw.v5i2.8458.

Al-Ansari, Y. D. Y. (2014). *Innovation practices as a path to business growth performance: A study of small and medium sized firms in the emerging UAE market*.

Albort-Morant, G., Leal-Millán, A., & Cepeda-Carrión, G. (2016). The antecedents of green innovation performance: A model of learning and capabilities. *Journal of Business Research*. https://doi.org/10.1016/j.jbusres.2016.04.052.

Alvarez-Herranz, A., Balsalobre-Lorente, D., Shahbaz, M., & Cantos, J. M. (2017). Energy innovation and renewable energy consumption in the correction of air pollution levels. *Energy Policy*. https://doi.org/10.1016/j.enpol.2017.03.009.

Bellotti, B., & Rochecouste, J. F. (2014). The development of conservation agriculture in Australia—farmers as innovators. *International Soil and Water Conservation Research*. https://doi.org/10.1016/S2095-6339(15)30011-3.

Berre, D., Vayssières, J., Boussemart, J. P., Leleu, H., Tillard, E., & Lecomte, P. (2015). A methodology to explore the determinants of eco-efficiency by combining an agronomic whole-farm simulation model and efficient frontier. *Environmental Modelling and Software*. https://doi.org/10.1016/j.envsoft.2015.05.008.

Bibri, S. E. (2018). The IoT for smart sustainable cities of the future: An analytical framework for sensor-based big data applications for environmental sustainability. *Sustainable Cities and Society*. https://doi.org/10.1016/j.scs.2017.12.034.

Boons, F., & Lüdeke-Freund, F. (2013). Business models for sustainable innovation: State-of-the-art and steps towards a research agenda. *Journal of Cleaner Production*. https://doi.org/10.1016/j.jclepro.2012.07.007.

Bossle, M. B., Dutra De Barcellos, M., Vieira, L. M., & Sauvée, L. (2016). The drivers for adoption of eco-innovation. *Journal of Cleaner Production*. https://doi.org/10.1016/j.jclepro.2015.11.033.

Brundtland, G. H. (1987). Our common future (Brundtland report). In *United Nations Commission*. https://doi.org/10.1080/07488008808408783.

Can, U., & Alatas, B. (2017). Big social network data and sustainable economic development. *Sustainability (Switzerland)*. https://doi.org/10.3390/su9112027.

Canonico, P., Consiglio, S., Iacono, M. P., Mercurio, L., & Berni, A. (2014). Different organizational models and roles in smart city systems. *Ifkad 2014: 9th International Forum on Knowledge Asset Dynamics*.

Chalvatzis, K. J., Malekpoor, H., Mishra, N., Lettice, F., & Choudhary, S. (2019). Sustainable resource allocation for power generation: The role of big data in enabling interindustry architectural innovation. *Technological Forecasting and Social Change*. https://doi.org/10.1016/j.techfore.2018.04.031.

Chang, C. H. (2011). The influence of corporate environmental ethics on competitive advantage: The mediation role of green innovation. *Journal of Business Ethics*. https://doi.org/10.1007/s10551-011-0914-x.

Cheng, C. C. J., Yang, C. L., & Sheu, C. (2014). The link between eco-innovation and business performance: A Taiwanese industry context. *Journal of Cleaner Production*. https://doi.org/10.1016/j.jclepro.2013.09.050.

Coad, A., de Haan, P., & Woersdorfer, J. S. (2009). Consumer support for environmental policies: An application to purchases of green cars. *Ecological Economics*. https://doi.org/10.1016/j.ecolecon.2009.01.015.

Cuerva, M. C., Triguero-Cano, Á., & Córcoles, D. (2014). Drivers of green and non-green innovation: Empirical evidence in low-tech SMEs. *Journal of Cleaner Production*. https://doi.org/10.1016/j.jclepro.2013.10.049.

De Gennaro, M., Paffumi, E., & Martini, G. (2016). Big data for supporting low-carbon road transport policies in Europe: Applications, challenges and opportunities. *Big Data Research*. https://doi.org/10.1016/j.bdr.2016.04.003.

Deng, X., & Gibson, J. (2019). Improving eco-efficiency for the sustainable agricultural production: A case study in Shandong, China. *Technological Forecasting and Social Change*. https://doi.org/10.1016/j.techfore.2018.01.027.

Dogliotti, S., García, M. C., Peluffo, S., Dieste, J. P., Pedemonte, A. J., Bacigalupe, G. F., … Rossing, W. A. H. (2014). Co-innovation of family farm systems: A systems approach to sustainable agriculture. *Agricultural Systems*. https://doi.org/10.1016/j.agsy.2013.02.009.

Elkington, J. (2004). Enter the triple bottom line. In A. Henriques, & J. Richardson (Eds.), *The triple bottom line: Does it all add up?* London, UK: Earthscan.

Elkington, John (2013). Enter the triple bottom line. In *The triple bottom line: Does it all add up*. https://doi.org/10.4324/9781849773348.

Fei, J., Wang, Y., Yang, Y., Chen, S., & Zhi, Q. (2016). Towards eco-city: The role of green innovation. *Energy Procedia*. https://doi.org/10.1016/j.egypro.2016.12.029.

Fussler, C. (1996). Driving eco-innovation: A breakthrough discipline for innovation and sustainability. In *Pitman publishing*. https://doi.org/10.1002/(SICI)1099-0836(199711)6:5<297::AID-BSE128>3.0.CO;2-R.

Garlapati, V. K. (2016). E-waste in India and developed countries: Management, recycling, business and biotechnological initiatives. *Renewable and Sustainable Energy Reviews*. https://doi.org/10.1016/j.rser.2015.10.106.

Greene, W. H. (2008). The econometric approach to efficiency analysis. In *The measurement of productive efficiency and productivity change*. https://doi.org/10.1093/acprof:oso/9780195183528.003.0002.

Hampton, S. E., Strasser, C. A., Tewksbury, J. J., Gram, W. K., Budden, A. E., Batcheller, A. L., … Porter, J. H. (2013). Big data and the future of ecology. *Frontiers in Ecology and the Environment*. https://doi.org/10.1890/120103.

Hassan, G. E., Salah, A. H., Fath, H., Elhelw, M., Hassan, A., & Saqr, K. M. (2016). Optimum operational performance of a new stand-alone agricultural greenhouse with integrated-TPV solar panels. *Solar Energy*. https://doi.org/10.1016/j.solener.2016.07.017.

Hojnik, J., & Ruzzier, M. (2016). The driving forces of process eco-innovation and its impact on performance: Insights from Slovenia. *Journal of Cleaner Production*. https://doi.org/10.1016/j.jclepro.2016.06.002.

Huang, Y., Chen, Z. X., Yu, T., Huang, X. Z., & Gu, X. F. (2018). Agricultural remote sensing big data: Management and applications. *Journal of Integrative Agriculture*. https://doi.org/10.1016/S2095-3119(17)61859-8.

Iraldo, F., Testa, F., & Frey, M. (2009). Is an environmental management system able to influence environmental and competitive performance? The case of the eco-management and audit scheme (EMAS) in the European Union. *Journal of Cleaner Production*. https://doi.org/10.1016/j.jclepro.2009.05.013.

Junior, C. H., Oliveira, T., & Yanaze, M. (2019). The adoption stages (evaluation, adoption, and routinisation) of ERP systems with business analytics functionality in the context of farms. *Computers and Electronics in Agriculture*. https://doi.org/10.1016/j.compag.2018.11.028.

Kammerer, D. (2009). The effects of customer benefit and regulation on environmental product innovation. Empirical evidence from appliance manufacturers in Germany. *Ecological Economics*. https://doi.org/10.1016/j.ecolecon.2009.02.016.

Kemp, R., & Pearson, P. (2007). *Final report MEI project about measuring eco-innovation*. Maastricht: UM Merit.

Ketata, I., Sofka, W., & Grimpe, C. (2015). The role of internal capabilities and firms' environment for sustainable innovation: Evidence for Germany. *R and D Management*. https://doi.org/10.1111/radm.12052.

Koskela-Huotari, K., Edvardsson, B., Jonas, J. M., Sörhammar, D., & Witell, L. (2016). Innovation in service ecosystems-breaking, making, and maintaining institutionalized rules of resource integration. *Journal of Business Research*. https://doi.org/10.1016/j.jbusres.2016.02.029.

Kumar, R., & Chandrakar, R. (2012). Overview of green supply chain management: Operation and environmental impact at different stages of the supply chain. *International Journal of Engineering and Advanced Technology*.

Kunapatarawong, R., & Martínez-Ros, E. (2016). Towards green growth: How does green innovation affect employment? *Research Policy*. https://doi.org/10.1016/j.respol.2016.03.013.

Lai, X., & Shi, Q. (2020). *Green and low-carbon technology innovations*.

Lawal, A. Q. T., Ninsiima, E., Odebiyib, O. S., Hassan, A. S., Oyagbola, I. A., Onu, P., & Yusuf, D. A. (2019). Effect of unburnt rice husk on the properties of concrete. *Procedia Manufacturing, 35*, 635–640. https://doi.org/10.1016/j.promfg.2019.06.006.

Li, K., & Lin, B. (2016). Impact of energy technology patents in China: Evidence from a panel cointegration and error correction model. *Energy Policy.* https://doi.org/10.1016/j.enpol.2015.11.034.

Lindman, Å., & Söderholm, P. (2016). Wind energy and green economy in Europe: Measuring policy-induced innovation using patent data. *Applied Energy.* https://doi.org/10.1016/j.apenergy.2015.10.128.

Lorente, D. B., & Álvarez-Herranz, A. (2016). Economic growth and energy regulation in the environmental Kuznets curve. *Environmental Science and Pollution Research.* https://doi.org/10.1007/s11356-016-6773-3.

Luhar, S., & Luhar, I. (2019). Potential application of E-wastes in construction industry: A review. *Construction and Building Materials.* https://doi.org/10.1016/j.conbuildmat.2019.01.080.

Mamede, P., & Gomes, C. F. (2014). Corporate sustainability measurement in service organizations: A case study from Portugal. *Environmental Quality Management.* https://doi.org/10.1002/tqem.21370.

Matos, S., & Silvestre, B. S. (2013). Managing stakeholder relations when developing sustainable business models: The case of the Brazilian energy sector. *Journal of Cleaner Production.* https://doi.org/10.1016/j.jclepro.2012.04.023.

McGorry, P. D., Ratheesh, A., & O'Donoghue, B. (2018). Early intervention—an implementation challenge for 21st century mental health care. *JAMA Psychiatry.* https://doi.org/10.1001/jamapsychiatry.2018.0621.

Meeusen, W., & van Den Broeck, J. (1977). Efficiency estimation from Cobb-Douglas production functions with composed error. *International Economic Review.* https://doi.org/10.2307/2525757.

Moldan, B., Janoušková, S., & Hák, T. (2012). How to understand and measure environmental sustainability: Indicators and targets. *Ecological Indicators.* https://doi.org/10.1016/j.ecolind.2011.04.033.

Montalvo, C. (2008). General wisdom concerning the factors affecting the adoption of cleaner technologies: A survey 1990–2007. *Journal of Cleaner Production.* https://doi.org/10.1016/j.jclepro.2007.10.002.

Morelli, J. (2011). Environmental sustainability: A definition for environmental professionals. *Journal of Environmental Sustainability.* https://doi.org/10.14448/jes.01.0002.

Morota, G., Ventura, R. V., Silva, F. F., Koyama, M., & Fernando, S. C. (2018). Big data analytics and precision animal agriculture symposium: Machine learning and data mining advance predictive big data analysis in precision animal agriculture. *Journal of Animal Science.* https://doi.org/10.1093/jas/sky014.

Mor, R. S., Singh, S., Bhardwaj, A., & Singh, L. (2015). Technological implications of supply chain practices in agri-food sector-A review. *International Journal of Supply and Operations Management.*

Mustapha, M. A., Manan, Z. A., & Wan Alwi, S. R. (2017). Sustainable Green Management System (SGMS) – an integrated approach towards organisational sustainability. *Journal of Cleaner Production.* https://doi.org/10.1016/j.jclepro.2016.06.033.

Niero, M., Olsen, S. I., & Laurent, A. (2018). Renewable energy and carbon management in the Cradle-to-Cradle certification: Limitations and opportunities. *Journal of Industrial Ecology.* https://doi.org/10.1111/jiec.12594.

OECD. (2010). Interim report of the green growth strategy: Implementing our commitment for a sustainable future. *Strategy.*

Oltra, V., & Saint Jean, M. (2009). Sectoral systems of environmental innovation: An application to the French automotive industry. *Technological Forecasting and Social Change.* https://doi.org/10.1016/j.techfore.2008.03.025.

Onu, Peter, & Mbohwa, C. (2018a). Future energy systems and sustainable emission control: Africa in perspective. *Proceedings of the International Conference on Industrial Engineering and Operations Management.*

Onu, Peter, & Mbohwa, C. (2018b). Green supply chain management and sustainable industrial practices: Bridging the gap. *Proceedings of the International Conference on Industrial Engineering and Operations Management*, 786–792 (Washington DC).

Onu, P., & Mbohwa, C. (2019a). A sustainable industrial development approach: Enterprice risk management in view. *Journal of Physics: Conference Series, 1378*(2). https://doi.org/10.1088/1742-6596/1378/2/022094.

Onu, P., & Mbohwa, C. (2019b). Sustainable production: New thinking for SMEs. *Journal of Physics: Conference Series, 1378*(2). https://doi.org/10.1088/1742-6596/1378/2/022072.

Onu, P., & Mbohwa, C. (2019c). Sustainable supply chain management: Impact of practice on manufacturing and industry development. *Journal of Physics: Conference Series, 1378*(2). https://doi.org/10.1088/1742-6596/1378/2/022073.

Pan, X., Ai, B., Li, C., Pan, X., & Yan, Y. (2019). Dynamic relationship among environmental regulation, technological innovation and energy efficiency based on large scale provincial panel data in China. *Technological Forecasting and Social Change.* https://doi.org/10.1016/j.techfore.2017.12.012.

Pearson, P. J. G., & Foxon, T. J. (2012). A low carbon industrial revolution? Insights and challenges from past technological and economic transformations. *Energy Policy.* https://doi.org/10.1016/j.enpol.2012.07.061.

Peter, O., & Mbohwa, C. (2018a). Correlation between future energy systems and industrial revolutions. *Proceedings of the International Conference on Industrial Engineering and Operations Management*, 1953–1961 (Pretoria/Johannesburg).

Peter, O., & Mbohwa, C. (2018b). The interlink between sustainable supply chain management and technology development in industry. *Proceedings of the International Conference on Industrial Engineering and Operations Management*, 425–430 (Pretoria/Johannesburg).

Peter, O., & Mbohwa, C. (2019). Industrial energy conservation initiative and prospect for sustainable manufacturing. *Procedia Manufacturing, 35*, 546–551. https://doi.org/10.1016/j.promfg.2019.05.077.

Reinhard, S., Lovell, C. A. K., & Thijssen, G. (1999). Econometric estimation of technical and environmental efficiency: An application to Dutch dairy farms. *American Journal of Agricultural Economics.* https://doi.org/10.2307/1244449.

Rimos, S., Hoadley, A. F. A., & Brennan, D. J. (2014). Environmental consequence analysis for resource depletion. *Process Safety and Environmental Protection.* https://doi.org/10.1016/j.psep.2013.06.001.

Robaina-Alves, M., Moutinho, V., & MacEdo, P. (2015). A new frontier approach to model the eco-efficiency in European countries. *Journal of Cleaner Production.* https://doi.org/10.1016/j.jclepro.2015.01.038.

Ruan, J., Wang, Y., Chan, F. T. S., Hu, X., Zhao, M., Zhu, F., ... Lin, F. (2019). A life cycle framework of green IoT-based agriculture and its finance, operation, and management issues. *IEEE Communications Magazine.* https://doi.org/10.1109/MCOM.2019.1800332.

Schiederig, T., Tietze, F., & Herstatt, C. (2012). Green innovation in technology and innovation management – an exploratory literature review. *R and D Management*. https://doi.org/10.1111/j.1467-9310.2011.00672.x.

Shen, B., Choi, T. M., & Chan, H. L. (2019). Selling green first or not? A Bayesian analysis with service levels and environmental impact considerations in the big data era. *Technological Forecasting and Social Change*. https://doi.org/10.1016/j.techfore.2017.09.003.

Smol, M., Kulczycka, J., & Avdiushchenko, A. (2017). Circular economy indicators in relation to eco-innovation in European regions. *Clean Technologies and Environmental Policy*. https://doi.org/10.1007/s10098-016-1323-8.

Söderholm, K., Bergquist, A. K., & Söderholm, P. (2017). The transition to chlorine free pulp revisited: Nordic heterogeneity in environmental regulation and R&D collaboration. *Journal of Cleaner Production*. https://doi.org/10.1016/j.jclepro.2017.07.190.

Song, M., An, Q., Zhang, W., Wang, Z., & Wu, J. (2012). Environmental efficiency evaluation based on data envelopment analysis: A review. *Renewable and Sustainable Energy Reviews*. https://doi.org/10.1016/j.rser.2012.04.052.

Song, M., Wang, S., & Liu, Q. (2013). Environmental efficiency evaluation considering the maximization of desirable outputs and its application. *Mathematical and Computer Modelling, 58*(5–6), 1110–1116. https://doi.org/10.1016/j.mcm.2011.12.043.

Song, M. L., Cao, S. P., & Wang, S. H. (2019). The impact of knowledge trade on sustainable development and environment-biased technical progress. *Technological Forecasting and Social Change*. https://doi.org/10.1016/j.techfore.2018.02.017.

Song, M., Fisher, R., & Kwoh, Y. (2019). Technological challenges of green innovation and sustainable resource management with large scale data. *Technological Forecasting and Social Change*. https://doi.org/10.1016/j.techfore.2018.07.055.

Sousa-Zomer, T. T., & Cauchick Miguel, P. A. (2018). Sustainable business models as an innovation strategy in the water sector: An empirical investigation of a sustainable product-service system. *Journal of Cleaner Production*. https://doi.org/10.1016/j.jclepro.2016.07.063.

Tamayo-Orbegozo, U., Vicente-Molina, M. A., & Villarreal-Larrinaga, O. (2017). Eco-innovation strategic model. A multiple-case study from a highly eco-innovative European region. *Journal of Cleaner Production*. https://doi.org/10.1016/j.jclepro.2016.11.174.

Tansel, B. (2017). From electronic consumer products to E-wastes: Global outlook, waste quantities, recycling challenges. *Environment International*. https://doi.org/10.1016/j.envint.2016.10.002.

Tien, J. M. (2013). Big data: Unleashing information. *Journal of Systems Science and Systems Engineering*. https://doi.org/10.1007/s11518-013-5219-4.

United States Environmental Protection Agency. (2014). Air quality index: A guide to air quality and your health. In *Epa*. https://doi.org/10.1023/A:1020370119096.

Viaggi, D. (2015). Research and innovation in agriculture: Beyond productivity? *Bio-Based and Applied Economics*. https://doi.org/10.13128/BAE-17555.

Wagner, M. (2015). A European perspective on country moderation effects: Environmental management systems and sustainability-related human resource benefits. *Journal of World Business*. https://doi.org/10.1016/j.jwb.2014.08.005.

Wang, Z., Bao, Y., Wen, Z., & Tan, Q. (2016). Analysis of relationship between Beijing's environment and development based on environmental Kuznets curve. *Ecological Indicators*. https://doi.org/10.1016/j.ecolind.2016.02.045.

Wang, S. H., & Song, M. L. (2014). Review of hidden carbon emissions, trade, and labor income share in China, 2001–2011. *Energy Policy*. https://doi.org/10.1016/j.enpol.2014.08.038.

Weng, H. H. R., Chen, J. S., & Chen, P. C. (2015). Effects of green innovation on environmental and corporate performance: A stakeholder perspective. *Sustainability (Switzerland)*. https://doi.org/10.3390/su7054997.

Wiseman, J. (2018). The great energy transition of the 21st century: The 2050 zero-carbon world oration. *Energy Research and Social Science*. https://doi.org/10.1016/j.erss.2017.10.011.

World Bank. (2016). Agriculture, value added (% of GDP). *Data*. https://doi.org/10.1179/136485908x337463.

Wu, G. C. (2013). The influence of green supply chain integration and environmental uncertainty on green innovation in Taiwan's IT industry. *Supply Chain Management*. https://doi.org/10.1108/SCM-06-2012-0201.

Yang, M., Evans, S., Vladimirova, D., & Rana, P. (2017). Value uncaptured perspective for sustainable business model innovation. *Journal of Cleaner Production*. https://doi.org/10.1016/j.jclepro.2016.07.102.

Yim, S. H. L., Fung, J. C. H., & Lau, A. K. H. (2010). Use of high-resolution MM5/CALMET/CALPUFF system: SO_2 apportionment to air quality in Hong Kong. *Atmospheric Environment*. https://doi.org/10.1016/j.atmosenv.2010.08.037.

Yu, W., Chavez, R., Feng, M., & Wiengarten, F. (2014). Integrated green supply chain management and operational performance. *Supply Chain Management: An International Journal*. https://doi.org/10.1108/SCM-07-2013-0225.

Yun, S., & Lee, J. (2015). Advancing societal readiness toward renewable energy system adoption with a socio-technical perspective. *Technological Forecasting and Social Change*. https://doi.org/10.1016/j.techfore.2015.01.016.

Yusuf, A. A., Peter, O., Hassanb, A. S., Tunji, A. L., Oyagbola, I. A., Mustafad, M. M., & Yusuf, D. A. (2019). Municipality solid waste management system for Mukono district. *Procedia Manufacturing, 35*, 613–622. https://doi.org/10.1016/j.promfg.2019.06.003.

Zeng, M., & Stratton, R. (2017). Sustainable operations management. In *The business student's guide to sustainable management (second edition): Principles and practice*. https://doi.org/10.9774/gleaf.9781783533213_13.

CHAPTER

9 Which way forward: agricultural waste management and the fourth industrial revolution appraisal

9.1 Introduction

Recently, interventions in agronomy have led to social, economic, and environmental impacts that are changing the global perspective on agricultural biodiversity (Altieri, 2018; DeLonge, Miles, & Carlisle, 2016). There is a need to investigate the opportunities, challenges, and implementation strategies of nascent innovations linked to resource conservation and waste management (Kamble, Gunasekaran, & Gawankar, 2018; Miranda, Ponce, Molina, & Wright, 2019). Also, embracing the new technological trends and the Industry 4.0 paradigm calls for understanding in risk assessment and management skills to guide sustainable agricultural practices. This is vital to ensure adequate food supply and production of essential clean, green, and bio-based materials, thus alleviating poverty and conserving our natural resources (Hidayatno, Destyanto, & Hulu, 2019). Innovation in agriculture combines timely access and application of information that is available to proffer solutions to threats linked to unmanaged wastes (Royal Society of London, 2009).

The advancement of ICT in developing countries is breaking barriers linked to poor business practices. It has proven to be an essential tool for providing sustainable advisory services to farmers, which has led to an increase in productivity through innovative initiatives and integrated strategies (Binswanger-Mkhize & Savastano, 2017; Muzari, Gatsi, & Muvhunzi, 2012; Schut et al., 2016). The drawback of knowledge deficiency relating to poor informatization and technology utilization is linked to the weak implementation strategy. The main focus of the chapter explores the changing trends of agricultural waste management practices, with emphasis on the nascent technology arrivals and development strategies. Different perspectives in line with the near-future technology actualizations have been discussed. The assumption that the advancement of sustainability implementation and agricultural development, viz., Industry 4.0 application in agriculture diversity is validated, as it stands to promote excellence in the near future.

Agricultural Waste Diversity and Sustainability Issues. https://doi.org/10.1016/B978-0-323-85402-3.00010-3

Technological innovations for sustainable agriculture

One of the challenges that prompt the increase in resource control is inspired by the lack of responsibility toward the use of highly pragmatic and appropriate approaches that consider the trade-off between economic growth and environment protection (McKenzie & Williams, 2015; Tomlinson, 2013; West et al., 2014). Also, due to agroindustrialization and the availability of means for which consumers may become aware of the activities involved that contribute to climate change (Manning, Baines, & Chadd, 2006). The emergence of new retailer reforms and government intervention/incentives, the strict requirements for quality agrifarm produce, and environmental safety consideration is justified, thus resulting in intensified pressure to deliver on sustainable practices and technological innovation in the agricultural value chain (Allouche, 2011; Shiferaw et al., 2013). This, in turn, will not only yield enhanced economic and sustainable outcomes but will also lay a foundation to earn a competitive advantage for all the stakeholders involved in the operational structure (Carter & Rogers, 2008).

Numerous researches have iterated the need for the agroindustry sector to focus on sustainability and technology integration on different levels and within the structures of agricultural development, i.e., upstream, midstream, and downstream. The sustainability of an entire food supply chain operations processes tenable for sustainable development according to Zeng et al. (2017), has been explored in China (Zeng, Chen, Xiao, & Zhou, 2017). Kirwan, Maye, and Brunori (2017) identified and analyzed various attributes that contribute to the complexities in sustainable performance evaluation of the food supply chain, whereas Mani, Gunasekaran, and Delgado (2018) explored the measures related to supplier social sustainability in emerging economies. Other researchers engaged the topic from the advanced information communication and integration perspective in the agroindustry. Accordingly, advances made in science and technology have been beneficial to agriculture, making it profitable while assuring sustainability (Long, Blok, & Coninx, 2016; Pierpaoli, Carli, Pignatti, & Canavari, 2013). Technology from their perspective has also reduced the burden on farmers in that sophisticated types of machinery now perform most jobs that ordinarily would have been carried out by farmers (Pivoto et al., 2018; Qu, Wang, Kang, & Liu, 2018).

The action plan is for businesses to redesign their strategic paths and operational measures that will support their productivity target and sustainability goals to overcome the impending challenges (Dyllick & Hockerts, 2002; Evans et al., 2017; Hubbard, 2009). New groups and institutions that are responsible for the promotion of sustainable technology innovation and agricultural development must spring up in all regions to affect all levels of knowledge and the business practices in resource conservation (Carayannis, Sindakis, & Walter, 2015; Hockerts, 2015). Use of energy-efficient equipment both for production and during packaging operations or recycling as a practice of sustainability implementation among other strategies can be adopted for environmental abatement and pollution reduction. Farmers in developing countries and Africa in perspective are expected to deliver standardized technologies to compete favorably and satisfy conditions for exportation to other countries. Hence, the mandate is to provide future scope through which enterprises can incorporate as part of their ongoing efforts toward sustainability (Searcy, 2016).

9.2 Advances in information communication and knowledge management for agrowaste reduction

It is greatly important that the development of science and technology goes beyond mediocre standards to overcome the negative impartation of agricultural activities on the environment. The idea of ecoliteracy (environment/ecology literacy) or awareness of the world community is a paradigm that leads to increased ecological consciousness of a society (Kim, Vaswani, Kang, Nam, & Lee, 2017; Muliana, Maryani, & Somantri, 2018). This can be achieved through learning provision at the various institutes of higher education where demonstration of the techniques and deployment of necessary innovative ideas are launched. Nugraha (2016) has stated that it is essential for environmental literacy to be primarily promoted, as it transforms people's understanding of the global ecological awareness through the different models or methods that aim to improve acceptance of the concept for sustainable impartation. As such, the primary reason for learning using a problem-solving model is to improve high-level thinking skills.

HD high-level thinking and share of competencies has proven to offer resolves, through the several characteristics, in which the various solutions, involving different criteria and encourages self-regulation, and requires sincere mental effort (Hidayat, Susilaningsih, & Kurniawan, 2018; Scherer & Tiemann, 2014). The importance of problem-solving competencies for students has been consistently mandated (Brundiers, Wiek, & Redman, 2010; Eseryel, Law, Ifenthaler, Ge, & Miller, 2013; Remington-Doucette, Connell, Armstrong, & Musgrove, 2013). Thus, learning with problem-solving in schools can help students further understand the main concepts and principles of applications to solve problems, especially to increase ecology literacy and promote both biodegradable waste management, and green strategies (Fünfgeld, 2015; Rauken, Mydske, & Winsvold, 2015).

9.3 Integrated farming and agriculture diversity

The worldwide threat on food security and the effects of anthropogenic activities linked to agricultural exploitation for food and production of materials is a common issue that continues to attract global attention in a bid for sustainable habitation of the earth (Lawal et al., 2019; Yusuf et al., 2019; Kuhad & Singh, 2013; Pattanaik, Pattnaik, Saxena, & Naik, 2019; Onu & Mbohwa, 2018c; World Bank, 2016). The evolution of modern technologies for agricultural development is celebrated worldwide. With the entry of digitization as a measure for increased productivity, output has improved upon the application of one form of advanced technological assisted process or another (Bendre, Thool, & Thool, 2016). The need to have a simplified but highly advanced approach of coordinating agricultural development strategies has been discussed and relates man, plants, animals, and mechanical devices to be interlinked for agriculture diversity (Bendre et al., 2016; Poulsen, McNab, Clayton, & Neff, 2015).

Sustainable agriculture pertains to the activities that facilitate the production and processing of food, feed, fiber, and green fuel resources, which result in significant reduction in hard labor, effectiveness in delivery, and conservation of natural resources. However, the application of new and improved agricultural strategies globally or development from what stemmed to be unsophisticated has transformed into multimodal interaction devices, thanks to advances in technologies (Aqeel-Ur-Rehman, Abbasi, Islam, & Shaikh, 2014; Qu, Zhu, Sarkis, Geng, & Zhong, 2013). This can reduce operational costs and contributes to the protection of viable lands from indiscriminate dumping, toxic emissions, and climate change mitigation (Logan & Rabaey, 2012). There have been noticeable advances and a variety of emerging technologies that have facilitated the practice of agricultural development and led to the intervention of the fourth industrial revolution concepts: precision agriculture, smart farms, drone farming, and GPS-based applications, among others. The following section focuses on the trends, approaches, systems, and uses of the various emerging technologies that are predominant to cause rapid change and development in agricultural production and waste management.

Agricultural development and the Industry 4.0 paradigm

The conceptualization of the industry 4.0 paradigm by the German scientists to optimize manufacturing in 2011, was birthed in 1999. As such, the idea which was finally penned on paper, 10 years after by Kevin Ashton (Ashton, 2009) in 2009 has become a globally aceepted concept which manufacturers can relate to in terms of productivity. Industry 4.0 aims to digitize, robotize, and automate human activities to ensure higher speed and efficiency production. Some of the visions of the initiative are to replace manual human labor by automating data collection, production sequence, operational procedures, and robotically delivering component manufactured products or services (Bonneau, Copigneaux, Probst, & Pedersen, 2017; Zambon, Cecchini, Egidi, Saporito, & Colantoni, 2019). Recently, the aspect and need to advance the development of Industry 4.0 for a restructuring of the educational system are linked to strategies that will facilitate agricultural development globally (Yahya, 2018). As such, the synergy between education and the state of affairs of activities surrounding the Industry 4.0 paradigm to promote agrowaste management can lead to a positive resolve toward the actualization of the sustainable development goals.

This is prognosis about the transformative change in the future and healthy competition between private and public enterprises, especially in the agriculture sector. The education system, no doubt, requires restructuring, particularly in the secondary/high schools and university levels, and in line with sustainable approaches to apply emerging technological skill set to improve livelihood. The emphasis is to rethink sustainable policies that will promulgate the application of new and emerging technologies and in the agroindustry. The call is, therefore, being made to the sub-Saharan African nations and her leaders to establish a feasible and strategic approach that will capitalize on the full potential of digitization and automation. In essence, this ensures resources availability and qualified staff to manage/operate the new systems whenever implemented.

Industry 4.0 approach, systems and implementation in agriculture

The inclusion of the Industry 4.0 components in agricultural development is critical to ensure improved operational activities. This will promote the adoption of advanced technologies, such as mobile internet, smart sensors, cloud computing, the Internet of things (IoT), 3D printing, and much more, as effectively as possible (Annosi, Brunetta, Monti, & Nat, 2019; Trivelli et al., 2019). The nature of the technology to be used, therefore, ought to permit sustainability across the agriculture supply chain, while fostering interconnection between the customers, retailers, and suppliers (Jørgensen, 2018). The agricultural industry need to approach the Industry 4.0 advancement with the perspective of not just wanting to digitize some sector (food production and processing, dairy farming, crop cultivation, etc.) but also contemplating how the operations can become smart and sustainable. The prospect nowadays is toward smart waste management and agriculture diversity as a functioning Industry 4.0-based agricultural development, which goes beyond the training of personnel. As such, it is essential to identify risks and opportunities to ensure a credible approach to agricultural development and biodegradable waste management progress (Tseng, Chiu, Chien, & Tan, 2019).

The emergence of IoT in the agricultural sector, and the chances for its application, would mean increased connectivity and information sharing, which can impact the supply chain and its resilience to overcome all nature of disruptions and wastes, leading to poor sustainability practices (Gacar, Aktas, & Ozdogan, 2017). Supply chains are a core area for organizations, wherein they are complex and often unpredictable, as they include four essential functions: sales, distribution, production, and procurement (Ikumapayi, Akinlabi, Onu, Akinlabi, & Agarana, 2019; Onu & Mbohwa, 2019a, 2019b, 2018a; Onu & Mbohwa, 2018b). Advanced mechanized farms around the world have automated their processes, which has contributed to an increase in the amount of digitized data both internally within an organization and externally through distributors, freight, and transportation providers.

Internet of things

IoT deals with the interaction of physical objects with virtual representations, linked to information about which may be freely shared. This information is collected from real-time sensor data on conditions, properties, or any other useful aspect relating to the objects (Aggarwal, Ashish, & Sheth, 2014). The architecture of IoT is composed of sensors, actuators, and controllers, and of course, the "things" that are embedded in all these mechanisms. This portrays the IoT as a fusion of various systems that work together to transmit, collect, and exchange data (Onu & Mbohwa, 2019c, 2019d). IoT allows for sensing of objects, and this can be done remotely across a network since it projects interaction between the physical world and the world of the computer. When this is done, it results in efficiency and precision, thereby conferring economic benefits (Gondchawar & Kawitkar, 2016).

In the agricultural sector, it was identified that the producers needed a more significant interaction between their equipment and the environment. This interaction will permit easier means to observe and identify aspects that can interrupt in the production operations, without the use of other specific equipment (Verdugo Cedeño, Papinniemi, Hannola, & Donoghue, 2018). Technologies in agricultural development are fast evolving and aim at easing the conduct of agricultural production globally (Stočes, Vaněk, Masner, & Pavlík, 2016). Most of these technologies are still being conceptualized in universities and laboratories, while others are residents in highly industrialized nations. In Africa, however, the system of agricultural production is crude and mostly relies on rainfall for the years of production. This seems inappropriate as there must be a concerted effort to alter artificially and make natural elements available all year round to ensure food security. An example is the artificial/induced rain in Dubai, UAE, to moderate the temperature and farming activities of nearby settlements.

Through the application of the "IoT" concept, environmental conditions can be monitored, such as temperature, humidity, light intensity, and soil moisture in a greenhouse arrangement (Tzounis, Katsoulas, Bartzanas, & Kittas, 2017; Verdouw, Wolfert, & Tekinerdogan, 2016). The IoT can also ensure proper tracking of agricultural products from the farm to the final consumers. We learn from the literature that, with an Internet-based architecture, it was possible to monitor the operation of a tractor: to predict, failures, and maintenance needs without complications (Porkodi, Yuvaraj, Mohammed, Sivaram, & Manikandan, 2018). As such, data are processed and accompanied in software to offer customers beyond the sale of tractors, scheduled maintenance operations. This implies that the data concerning the equipment condition, production rates, energy consumption, and maintenance rates, when managed properly, will become transform agricultural operations for good (Fang et al., 2014). Hence, it will become a factor to generate knowledge, which also supports the decision-making processes.

Big data

The concept of understanding and utilization of data is essential in agricultural production, especially in agribusiness. As described in the works of Carbonell (2016), big data can be useful in forecasting to improve the efficiency of agricultural production (Liakos, Busato, Moshou, Pearson, & Bochtis, 2018). Importantly, data learning and management must be well founded to be assertive of generating a competitive advantage that is visible through the interconnection of production (Fleming, Jakku, Lim-Camacho, Taylor, & Thorburn, 2018; Kamilaris, Kartakoullis, & Prenafeta-

Boldú, 2017; Majumdar, Naraseeyappa, & Ankalaki, 2017). There are beneficial interactions with corporations when new technological systems are available to predict the occurrence of failures and indicate the part needed to perform maintenance services, or, if possible, to avoid production stoppages by monitoring the operating conditions of the machine to react faster to problems (Bertot, Gorham, Jaeger, Sarin, & Choi, 2014; Nagy, Oláh, Erdei, Máté, & Popp, 2018). The urgency of the process where quick decisions are made from data analysis and the capability to overcome the challenges of storing a large volume of data has become inevitable to progressive business owners (Márquez & Lev, 2016; Wigan & Clarke, 2013).

Big data analytics creates a favorable ground for the collection of greater perceptions about the operations that support automation, leading to resolutions that provide benefits to the organizations (Akoum & Mahjoub, 2013). Many organizations use the big data features to process data generated primarily from the IoT application. However, such data must be transformed into knowledge. Existing and emerging enterprises that desire to have competitive advantages and grow smart businesses must use these technologies appropriately to achieve the desired goal (Arass, Tikito, & Souissi, 2018). An example of a solar fish farm, where IoT technologies are used to guide the operations and monitoring of a solar pumping system for a smart farm scenario, has been reported to be highly effective (Sirisamphanwong, Wongthai, & Ngoenmeesri, 2019). The technology can also assist farmers to make decisions concerning the location of their enterprises, as it impacts productivity and availability of labor to manage the farming activities.

Blockchain

Blockchain is a distributed data structure that is replicated and shared among the members of a network (Witte, 2017); it keeps a chronologically growing list (chain) of records (blocks) secure from tampering and revision. The use of blockchain technology allows for a more transparent supply chain of goods or services. Each block in the chain can enable a stakeholder to control information through secure, auditable, and immutable records (Muzammal, Qu, & Nasrulin, 2019; Wood, 2014). The most crucial area where the blockchain helps in is the guarantee of validity and accountability of transactions, such that it gives businesses the ability to view the entire supply chain (Ge et al., 2017). When each activity moves along the supply chain, all parties on the network are linked together through the private blockchain. This brings transparency and validity and reduces the risk between all parties (Kim, Hilton, Burks, & Reyes, 2019; Papa, 2017). Blockchain proves to be a convenient, private, and simplified mode of payment (Nakamoto, 2008); however, the technology is gradually being integrated and diversified to smart contracts, especially in the present era of digitization. A decentralized blockchain is more profound and widely adopted by businesses, including the agricultural sector, which operates about a network that is not centrally controlled (Kim & Laskowski, 2018; Tošić, Vičič, & Mrissa, 2019).

Hence, the flow of information occurs between various stakeholders, from the consumers to the retailer, then the wholesaler, to transportation and freight providers to the supplier, and finally to the resource provider. These transactions occur through multiple systems, and the passing of accurate information between them depends on the trust from one stakeholder to another. There is a risk of data being incorrect or altered along the way of the supply chain, where there is also a lack of traceability (Caro, Ali, Vecchio, & Giaffreda, 2018). More so, there is a close relationship between agriculture and food supply chain, since the end products of agricultural production are utilized as inputs in the supply (Schipmann & Qaim, 2011).

Furthermore, it offers a remarkable benefit in agriculture in that it can provide a convenient, safe, and secure way to carry out transactions among different untrusted parties. This is in line with the findings of Lin et al. (2017), who stated that transactions are key in the agriculture and food supply chain even when numerous players are involved, starting from the primary stage of production to the tertiary stage. In developing countries, blockchain can be a support to SME farmers. This agrees with authors (Tripoli & Schmidhuber, 2018) in their remark that blockchain could include financing and insurance cover for rural farmers to ease agriculture diversity, hence promoting efficiency, transparency, and traceability during transactions.

Intelligent farming innovations

The intelligent farming innovation uses dedicated sensor and antennas/tracking mechanism, global positioning system (GPS), robotics, etc., to mitigate agricultural resources management (Li, Guo, & Gao, 2015; Shekhar, Colletti, Krintz, & Muñoz-arriola, 2017; Sivamani, Bae, & Cho, 2013). The use of sensors helps in generating the data that are used to manage the operations, while working to solve problems remotely and immediately. In essence, users can record reduced costs of management, reduced environmental impact (as it optimizes waste generation), safety, and excellence in the food preservation and farming dispositions (Bogatinoska, Malekian, Trengoska, & Nyako, 2016).

The Internet serves as a network for process control, which includes the use of cell phones, for example, that functions as the system interface. As such, antennas can be used to meet the demands of security and radar tracking, as well as to offer customers services such as Wi-Fi and live television broadcast of farming operations (cattle tracking, trespasser, and visual performance of crops, etc.). The successful creation of these virtual models has made it easy to evaluate the performance improvement in each location. With the adoption of GPS, generic radio communication techniques can be applied to mobile telephones or other networked systems. This enables remote monitoring of farms and machinery and maps out gracing fields for livestock in a network protocol for precision farming (Srbinovska, Gavrovski, Dimcev, Krkoleva, & Borozan, 2015; Suma, Samson, Saranya, Shanmugapriya, & Subhashri, 2017).

The adoption of other smart and intelligent machines that apply inputs in effective ways, for the development of small, durable, and smart machines, can address waste reduction, environmental protection, and guarantee sustainable food production as well. McKinion and Lemmon in 1985 first proposed the use of artificial intelligence (AI) to be applied in general crop management in their paper "Expert

Systems for Agriculture" (McKinion & Lemmon, 1985). Its operation is paramount in pest management, disease management, agricultural product monitoring, storage control, soil and irrigation management, weed management, yield prediction, etc. The former has brought about innovation in robotic assistance in agricultural development. Robotics applies to a variety of functions in the agricultural sector, which cuts across cultivation and harvests, since they can be used in the livestock sector for operations ranging from milking, cow dung disposal, cleaning of barns, field fencing, etc. (Bloss, 2014; Drewry, Shutske, Trechter, Luck, & Pitman, 2019; Tsiropoulos, Fountas, Liakos, & Tekin, 2013).

9.4 Conclusions and perspectives

The world is always changing, and with the recent advances through the various technological eras: first, second, third, and fourth, what lies ahead is unclear. The 21st century has witnessed formidable technological evolution, which holds promise to transform livelihood and change operations as we know it. While there is the perspective of complex systems, emerging to replace the traditional ways, and the speculations of the unknown due to different challenges (operational cost, technical capability, technology availability, and sustainability assurance of the systems, etc.), there is no turning back now. The development of standard protocols to communicate and transmit knowledge base on nascent technological changes must reach the three levels of stakeholder (upstream, midstream, and downstream). Beyond the education and knowledge management criteria, it is essential to investigate the specific technologies that will guarantee performance improvement and precision agriculture.

References

Aggarwal, C. C., Ashish, N., & Sheth, A. (2014). The internet of things: A survey from the data-centric perspective. In *Managing and mining sensor data*. https://doi.org/10.1007/978-1-4614-6309-2_12.

Akoum, M., & Mahjoub, A. (2013). A unified framework for implementing business intelligence, real-time operational intelligence and big data analytics for upstream oil industry operators. In *Society of petroleum engineers - SPE intelligent energy international 2013: Realising the full asset value*. https://doi.org/10.2118/167410-ms.

Allouche, J. (2011). The sustainability and resilience of global water and food systems: Political analysis of the interplay between security, resource scarcity, political systems and global trade. *Food Policy*. https://doi.org/10.1016/j.foodpol.2010.11.013.

Altieri, M. A. (2018). Agroecology: The science of sustainable agriculture. In *Agroecology: The science of sustainable agriculture* (2nd ed.). https://doi.org/10.1201/9780429495465.

Annosi, M. C., Brunetta, F., Monti, A., & Nat, F. (2019). Is the trend your friend? An analysis of technology 4.0 investment decisions in agricultural SMEs. *Computers in Industry*. https://doi.org/10.1016/j.compind.2019.04.003.

Aqeel-Ur-Rehman, Abbasi, A. Z., Islam, N., & Shaikh, Z. A. (2014). A review of wireless sensors and networks' applications in agriculture. *Computer Standards and Interfaces*. https://doi.org/10.1016/j.csi.2011.03.004.

Arass, M. El, Tikito, I., & Souissi, N. (2018). An audit framework for data lifecycles in a big data context. In *2018 international conference on selected topics in Mobile and Wireless Networking, MoWNeT 2018*. https://doi.org/10.1109/MoWNet.2018.8428883.

Ashton, K. (2009). In the real world, things matter more than ideas. *RFID Journal*. http://www.rfidjournal.com/articles/view?4986.

Bendre, M. R., Thool, R. C., & Thool, V. R. (2016). Big data in precision agriculture: Weather forecasting for future farming. In *Proceedings on 2015 1st international conference on Next Generation Computing Technologies, NGCT 2015*. https://doi.org/10.1109/NGCT.2015.7375220.

Bertot, J. C., Gorham, U., Jaeger, P. T., Sarin, L. C., & Choi, H. (2014). Big data, open government and e-government: Issues, policies and recommendations. *Information Polity*. https://doi.org/10.3233/IP-140328.

Binswanger-Mkhize, H. P., & Savastano, S. (2017). Agricultural intensification: The status in six African countries. *Food Policy*. https://doi.org/10.1016/j.foodpol.2016.09.021.

Bloss, R. (2014). Robot innovation brings to agriculture efficiency, safety, labor savings and accuracy by plowing, milking, harvesting, crop tending/picking and monitoring. *Industrial Robot*. https://doi.org/10.1108/IR-08-2014-0382.

Bogatinoska, D. C., Malekian, R., Trengoska, J., & Nyako, W. A. (2016). Advanced sensing and internet of things in smart cities. In *2016 39th international convention on information and communication technology, electronics and microelectronics, MIPRO 2016 - proceedings*. https://doi.org/10.1109/MIPRO.2016.7522218.

Bonneau, V., Copigneaux, B., Probst, L., & Pedersen, B. (2017). Industry 4.0 in agriculture: Focus onIoT aspects. *Digital Transformation Monitor, 1*, 1–5.

Brundiers, K., Wiek, A., & Redman, C. L. (2010). Real-world learning opportunities in sustainability: From classroom into the real world. *International Journal of Sustainability in Higher Education*. https://doi.org/10.1108/14676371011077540.

Carayannis, E. G., Sindakis, S., & Walter, C. (2015). Business model innovation as lever of organizational sustainability. *The Journal of Technology Transfer*. https://doi.org/10.1007/s10961-013-9330-y.

Carbonell, I. M. (2016). The ethics of big data in big agriculture. *Internet Policy Review*. https://doi.org/10.14763/2016.1.405.

Caro, M. P., Ali, M. S., Vecchio, M., & Giaffreda, R. (2018). Blockchain-based traceability in agri-food supply chain management: A practical implementation. In *2018 IoT vertical and topical summit on agriculture - Tuscany, IOT Tuscany 2018*. https://doi.org/10.1109/IOT-TUSCANY.2018.8373021.

Carter, C. R., & Rogers, D. S. (2008). A framework of sustainable supply chain management: Moving toward new theory. *International Journal of Physical Distribution and Logistics Management*. https://doi.org/10.1108/09600030810882816.

DeLonge, M. S., Miles, A., & Carlisle, L. (2016). Investing in the transition to sustainable agriculture. *Environmental Science and Policy*. https://doi.org/10.1016/j.envsci.2015.09.013.

Drewry, J. L., Shutske, J. M., Trechter, D., Luck, B. D., & Pitman, L. (2019). Assessment of digital technology adoption and access barriers among crop, dairy and livestock producers in Wisconsin. *Computers and Electronics in Agriculture*. https://doi.org/10.1016/j.compag.2019.104960.

Dyllick, T., & Hockerts, K. (2002). Beyond the business case for corporate sustainability. *Business Strategy and the Environment*. https://doi.org/10.1002/bse.323.

Eseryel, D., Law, V., Ifenthaler, D., Ge, X., & Miller, R. (2013). An investigation of the interrelationships between motivation, engagement, and complex problem solving in game-based learning. *Educational Technology and Society, 17*(1), 42–53.

Evans, S., Vladimirova, D., Holgado, M., Van Fossen, K., Yang, M., Silva, E. A., & Barlow, C. Y. (2017). Business model innovation for sustainability: Towards a unified perspective for creation of sustainable business models. *Business Strategy and the Environment*. https://doi.org/10.1002/bse.1939.

Fang, S., Xu, L. Da, Zhu, Y., Ahati, J., Pei, H., Yan, J., & Liu, Z. (2014). An integrated system for regional environmental monitoring and management based on internet of things. In *IEEE transactions on industrial informatics*. https://doi.org/10.1109/TII.2014.2302638.

Fleming, A., Jakku, E., Lim-Camacho, L., Taylor, B., & Thorburn, P. (2018). Is big data for big farming or for everyone? Perceptions in the Australian grains industry. *Agronomy for Sustainable Development*. https://doi.org/10.1007/s13593-018-0501-y.

Fünfgeld, H. (2015). Facilitating local climate change adaptation through transnational municipal networks. *Current Opinion in Environmental Sustainability*. https://doi.org/10.1016/j.cosust.2014.10.011.

Gacar, A., Aktas, H., & Ozdogan, B. (2017). Digital agriculture practices in the context of agriculture 4.0. *Pressacademia*. https://doi.org/10.17261/pressacademia.2017.448.

Ge, L., Brewster, C., Spek, J., Smeenk, A., Top, J., van Diepen, F., … de Ruyter de Wildt, M. (2017). *Blockchain for agriculture and food*. Wageningen University & Research. https://doi.org/10.18174/426747.

Gondchawar, N., & Kawitkar, P. R. S. (2016). IoT based smart agriculture. *International Journal of Advanced Research in Computer and Communication Engineering*. https://doi.org/10.17148/IJARCCE.2016.56188.

Hidayatno, A., Destyanto, A. R., & Hulu, C. A. (2019). Industry 4.0 technology implementation impact to industrial sustainable energy in Indonesia: A model conceptualization. *Energy Procedia*. https://doi.org/10.1016/j.egypro.2018.11.133.

Hidayat, T., Susilaningsih, E., & Kurniawan, C. (2018). The effectiveness of enrichment test instruments design to measure students' creative thinking skills and problem-solving. *Thinking Skills and Creativity*. https://doi.org/10.1016/j.tsc.2018.02.011.

Hockerts, K. (2015). A cognitive perspective on the business case for corporate sustainability. *Business Strategy and the Environment*. https://doi.org/10.1002/bse.1813.

Hubbard, G. (2009). Measuring organizational performance: Beyond the triple bottom line. *Business Strategy and the Environment*. https://doi.org/10.1002/bse.564.

Ikumapayi, O. M., Akinlabi, E. T., Onu, P., Akinlabi, S. A., & Agarana, M. C. (2019). A generalized model for automation cost estimating systems (ACES) for sustainable manufacturing. *Journal of Physics: Conference Series, 1378*(3). https://doi.org/10.1088/1742-6596/1378/3/032043.

Jørgensen, M. H. (2018). Agricultural field production in an 'industry 4.0' concept. *Agronomy Research*. https://doi.org/10.15159/AR.18.007.

Kamble, S. S., Gunasekaran, A., & Gawankar, S. A. (2018). Sustainable industry 4.0 framework: A systematic literature review identifying the current trends and future perspectives. *Process Safety and Environmental Protection*. https://doi.org/10.1016/j.psep.2018.05.009.

Kamilaris, A., Kartakoullis, A., & Prenafeta-Boldú, F. X. (2017). A review on the practice of big data analysis in agriculture. *Computers and Electronics in Agriculture*. https://doi.org/10.1016/j.compag.2017.09.037.

Kim, M., Hilton, B., Burks, Z., & Reyes, J. (2019). Integrating blockchain, smart contract-tokens, and IoT to design a food traceability solution. In *2018 IEEE 9th annual Information Technology, Electronics and Mobile Communication Conference, IEMCON 2018*. https://doi.org/10.1109/IEMCON.2018.8615007.

Kim, H., & Laskowski, M. (2018). Agriculture on the blockchain: Sustainable solutions for food, farmers, and financing. *SSRN Electronic Journal*. https://doi.org/10.2139/ssrn.3028164.

Kim, G. W., Vaswani, R. T., Kang, W., Nam, M., & Lee, D. (2017). Enhancing ecoliteracy through traditional ecological knowledge in proverbs. *Sustainability (Switzerland)*. https://doi.org/10.3390/su9071182.

Kirwan, J., Maye, D., & Brunori, G. (2017). Acknowledging complexity in food supply chains when assessing their performance and sustainability. *Journal of Rural Studies*. https://doi.org/10.1016/j.jrurstud.2017.03.008.

Kuhad, R. C., & Singh, A. (2013). Biotechnology for environmental management and resource recovery. In *Biotechnology for environmental management and resource recovery*. https://doi.org/10.1007/978-81-322-0876-1.

Lawal, A. Q. T., Ninsiima, E., Odebiyib, O. S., Hassan, A. S., Oyagbola, I. A., Onu, P., & Yusuf, D. A. (2019). Effect of unburnt rice husk on the properties of concrete. *Procedia Manufacturing, 35*, 635–640. https://doi.org/10.1016/j.promfg.2019.06.006.

Liakos, K. G., Busato, P., Moshou, D., Pearson, S., & Bochtis, D. (2018). Machine learning in agriculture: A review. *Sensors (Switzerland)*. https://doi.org/10.3390/s18082674.

Li, J., Guo, M., & Gao, L. (2015). Application and innovation strategy of agricultural internet of things. *Nongye Gongcheng Xuebao/Transactions of the Chinese Society of Agricultural Engineering*. https://doi.org/10.11975/j.issn.1002-6819.2015.z2.031.

Lin, Y.-P., Petway, J., Anthony, J., Mukhtar, H., Liao, S.-W., Chou, C.-F., & Ho, Y.-F. (2017). Blockchain: The evolutionary next step for ICT E-agriculture. *Environments*. https://doi.org/10.3390/environments4030050.

Logan, B. E., & Rabaey, K. (2012). Conversion of wastes into bioelectricity and chemicals by using microbial electrochemical technologies. *Science*. https://doi.org/10.1126/science.1217412.

Long, T. B., Blok, V., & Coninx, I. (2016). Barriers to the adoption and diffusion of technological innovations for climate-smart agriculture in Europe: Evidence from The Netherlands, France, Switzerland and Italy. *Journal of Cleaner Production*. https://doi.org/10.1016/j.jclepro.2015.06.044.

Majumdar, J., Naraseeyappa, S., & Ankalaki, S. (2017). Analysis of agriculture data using data mining techniques: Application of big data. *Journal of Big Data*. https://doi.org/10.1186/s40537-017-0077-4.

Mani, V., Gunasekaran, A., & Delgado, C. (2018). Enhancing supply chain performance through supplier social sustainability: An emerging economy perspective. *International Journal of Production Economics*. https://doi.org/10.1016/j.ijpe.2017.10.025.

Manning, L., Baines, R. N., & Chadd, S. A. (2006). Quality assurance models in the food supply chain. *British Food Journal*. https://doi.org/10.1108/00070700610644915.

Márquez, F. P. G., & Lev, B. (2016). Big data management. In *Big data management*. https://doi.org/10.1007/978-3-319-45498-6.

McKenzie, F. C., & Williams, J. (2015). Sustainable food production: Constraints, challenges and choices by 2050. *Food Security*. https://doi.org/10.1007/s12571-015-0441-1.

McKinion, J. M., & Lemmon, H. E. (1985). Expert systems for agriculture. *Computers and Electronics in Agriculture*. https://doi.org/10.1016/0168-1699(85)90004-3.

Miranda, J., Ponce, P., Molina, A., & Wright, P. (2019). Sensing, smart and sustainable technologies for Agri-Food 4.0. *Computers in Industry.* https://doi.org/10.1016/j.compind.2019.02.002.

Muliana, A., Maryani, E., & Somantri, L. (2018). Ecoliteracy level of student teachers (study toward students of Universitas Syiah Kuala Banda Aceh). In *IOP conference series: Earth and environmental science.* https://doi.org/10.1088/1755-1315/145/1/012061.

Muzammal, M., Qu, Q., & Nasrulin, B. (2019). Renovating blockchain with distributed databases: An open source system. *Future Generation Computer Systems.* https://doi.org/10.1016/j.future.2018.07.042.

Muzari, W., Gatsi, W., & Muvhunzi, S. (2012). The impacts of technology adoption on smallholder agricultural productivity in sub-Saharan Africa: A review. *Journal of Sustainable Development.* https://doi.org/10.5539/jsd.v5n8p69.

Nagy, J., Oláh, J., Erdei, E., Máté, D., & Popp, J. (2018). The role and impact of industry 4.0 and the internet of things on the business strategy of the value chain-the case of Hungary. *Sustainability (Switzerland).* https://doi.org/10.3390/su10103491.

Nakamoto, S. (2008). Bitcoin: A peer-to-peer electronic cash system. *Consulted,* 1–9.

Nugraha, R. G. (2016). Meningkatkan ecoliteracy siswa sd melalui metode field-trip kegiatan ekonomi pada mata pelajaran ilmu pengetahuan sosial. *Mimbar Sekolah Dasar.* https://doi.org/10.17509/mimbar-sd.v2i1.1322.

Onu, P., & Mbohwa, C. (2018a). Sustainable oil exploitation versus renewable energy initiatives: A review of the case of Uganda. In *Proceedings of the international conference on industrial engineering and operations management, 2018(SEP), 1008–1015.* IEOM Society.

Onu, P., & Mbohwa, C. (2018b). Green supply chain management and sustainable industrial practices: Bridging the gap. In *Proceedings of the international conference on industrial engineering and operations management, 786–792. Washington DC.*

Onu, P., & Mbohwa, C. (2018c). Correlation between future energy systems and industrial revolutions. In *Proceedings of the international conference on industrial engineering and operations management, 1953–1961. Pretoria/Johannesburg.*

Onu, P., & Mbohwa, C. (2019a). Sustainable production: New thinking for SMEs. *Journal of Physics: Conference Series, 1378*(2). https://doi.org/10.1088/1742-6596/1378/2/022072.

Onu, P., & Mbohwa, C. (2019b). Sustainable supply chain management: Impact of practice on manufacturing and industry development. *Journal of Physics: Conference Series, 1378*(2). https://doi.org/10.1088/1742-6596/1378/2/022073.

Onu, P., & Mbohwa, C. (2019c). *Cloud computing and IOT application : Current statuses and prospect for industrial development* (pp. 1–14).

Onu, P., & Mbohwa, C. (2019d). *New future for sustainability and industrial development : Success in blockchain, internet of production, and cloud computing technology* (pp. 1–12).

Papa, S. F. (2017). Use of blockchain technology in agribusiness: Transparency and monitoring in agricultural trade. https://doi.org/10.2991/msmi-17.2017.9.

Pattanaik, L., Pattnaik, F., Saxena, D. K., & Naik, S. N. (2019). Biofuels from agricultural wastes. In *Second and third generation of feedstocks.* https://doi.org/10.1016/b978-0-12-815162-4.00005-7.

Pierpaoli, E., Carli, G., Pignatti, E., & Canavari, M. (2013). Drivers of precision agriculture technologies adoption: A literature review. *Procedia Technology.* https://doi.org/10.1016/j.protcy.2013.11.010.

Pivoto, D., Waquil, P. D., Talamini, E., Finocchio, C. P. S., Dalla Corte, V. F., & de Vargas Mores, G. (2018). Scientific development of smart farming technologies and their application in Brazil. *Information Processing in Agriculture*. https://doi.org/10.1016/j.inpa.2017.12.002.

Porkodi, V., Yuvaraj, D., Mohammed, A. S., Sivaram, M., & Manikandan, V. (2018). IoT in agriculture. *Journal of Advanced Research in Dynamical and Control Systems*, (14), 1986–1991.

Poulsen, M. N., McNab, P. R., Clayton, M. L., & Neff, R. A. (2015). A systematic review of urban agriculture and food security impacts in low-income countries. *Food Policy*. https://doi.org/10.1016/j.foodpol.2015.07.002.

Qu, D., Wang, X., Kang, C., & Liu, Y. (2018). Promoting agricultural and rural modernization through application of information and communication technologies in China. *International Journal of Agricultural and Biological Engineering*. https://doi.org/10.25165/j.ijabe.20181106.4228.

Qu, Y., Zhu, Q., Sarkis, J., Geng, Y., & Zhong, Y. (2013). A review of developing an e-wastes collection system in Dalian, China. *Journal of Cleaner Production*. https://doi.org/10.1016/j.jclepro.2013.02.013.

Rauken, T., Mydske, P. K., & Winsvold, M. (2015). Mainstreaming climate change adaptation at the local level. *Local Environment*. https://doi.org/10.1080/13549839.2014.880412.

Remington-Doucette, S. M., Connell, K. Y. H., Armstrong, C. M., & Musgrove, S. L. (2013). Assessing sustainability education in a transdisciplinary undergraduate course focused on real-world problem solving: A case for disciplinary grounding. *International Journal of Sustainability in Higher Education*. https://doi.org/10.1108/IJSHE-01-2012-0001.

Royal Society of London. (2009). Reaping the benefits: Science and the sustainable intensification of global agriculture. In *Chemical engineer*. https://doi.org/10.2104/mbr08064.

Scherer, R., & Tiemann, R. (2014). Evidence on the effects of task interactivity and grade level on thinking skills involved in complex problem solving. *Thinking Skills and Creativity*. https://doi.org/10.1016/j.tsc.2013.10.003.

Schipmann, C., & Qaim, M. (2011). Supply chain differentiation, contract agriculture, and farmers' marketing preferences: The case of sweet pepper in Thailand. *Food Policy*. https://doi.org/10.1016/j.foodpol.2011.07.004.

Schut, M., van Asten, P., Okafor, C., Hicintuka, C., Mapatano, S., Nabahungu, N. L., … Vanlauwe, B. (2016). Sustainable intensification of agricultural systems in the Central African Highlands: The need for institutional innovation. *Agricultural Systems*. https://doi.org/10.1016/j.agsy.2016.03.005.

Searcy, C. (2016). Measuring enterprise sustainability. *Business Strategy and the Environment*. https://doi.org/10.1002/bse.1861.

Shekhar, S., Colletti, J., Krintz, C., & Muñoz-arriola, F. (2017). Intelligent infrastructure for smart Agriculture : An integrated food, energy and water system. *Computing Community Consortium Catalyst*, 1–8.

Shiferaw, B., Smale, M., Braun, H. J., Duveiller, E., Reynolds, M., & Muricho, G. (2013). Crops that feed the world 10. Past successes and future challenges to the role played by wheat in global food security. *Food Security*. https://doi.org/10.1007/s12571-013-0263-y.

Sirisamphanwong, C., Wongthai, W., & Ngoenmeesri, R. (2019). An approach to enhance a solar pumping system with cloud computing and internet of things for Thailand smart farming 4.0. *ICIC Express Letters, Part B: Applications*. https://doi.org/10.24507/icicelb.10.02.147.

Sivamani, S., Bae, N., & Cho, Y. (2013). A smart service model based on ubiquitous sensor networks using vertical farm ontology. *International Journal of Distributed Sensor Networks*. https://doi.org/10.1155/2013/161495.

Srbinovska, M., Gavrovski, C., Dimcev, V., Krkoleva, A., & Borozan, V. (2015). Environmental parameters monitoring in precision agriculture using wireless sensor networks. *Journal of Cleaner Production*. https://doi.org/10.1016/j.jclepro.2014.04.036.

Stočes, M., Vaněk, J., Masner, J., & Pavlík, J. (2016). Internet of things (IoT) in agriculture - selected aspects. *Agris On-Line Papers in Economics and Informatics*. https://doi.org/10.7160/aol.2016.080108.

Suma, N., Samson, S. R., Saranya, S., Shanmugapriya, G., & Subhashri, R. (2017). IOT based smart agriculture monitoring system. *International Journal on Recent and Innovation Trends in Computing and Communication*. https://doi.org/10.1109/ICRAECT.2017.52.

Tomlinson, I. (2013). Doubling food production to feed the 9 billion: A critical perspective on a key discourse of food security in the UK. *Journal of Rural Studies*. https://doi.org/10.1016/j.jrurstud.2011.09.001.

Tošić, A., Vičič, J., & Mrissa, M. (2019). *A blockchain-based decentralized self-balancing architecture for the web of things*. https://doi.org/10.1007/978-3-030-30278-8_34.

Tripoli, M., & Schmidhuber, J. (2018). Emerging opportunities for the application of blockchain in the agri-food industry agriculture. In *Food and agriculture organization of the United Nations*.

Trivelli, L., Apicella, A., Chiarello, F., Rana, R., Fantoni, G., & Tarabella, A. (2019). From precision agriculture to Industry 4.0: Unveiling technological connections in the agrifood sector. *British Food Journal*. https://doi.org/10.1108/BFJ-11-2018-0747.

Tseng, M. L., Chiu, A. S. F., Chien, C. F., & Tan, R. R. (2019). Pathways and barriers to circularity in food systems. *Resources, Conservation and Recycling*. https://doi.org/10.1016/j.resconrec.2019.01.015.

Tsiropoulos, Z., Fountas, S., Liakos, V., & Tekin, A. B. (2013). Web-based farm management information system for agricultural robots. In *EFITA-WCCA-CIGR conference "sustainable agriculture through ICT innovation."*.

Tzounis, A., Katsoulas, N., Bartzanas, T., & Kittas, C. (2017). Internet of Things in agriculture, recent advances and future challenges. *Biosystems Engineering*. https://doi.org/10.1016/j.biosystemseng.2017.09.007.

Verdouw, C., Wolfert, S., & Tekinerdogan, B. (2016). Internet of things in agriculture. *CAB Reviews: Perspectives in Agriculture, Veterinary Science, Nutrition and Natural Resources*. https://doi.org/10.1079/PAVSNNR201611035.

Verdugo Cedeño, J. M., Papinniemi, J., Hannola, L., & Donoghue, I. D. M. (2018). Developing smart services by internet of things in manufacturing business. *DEStech Transactions on Engineering and Technology Research*. https://doi.org/10.12783/dtetr/icpr2017/17680.

West, P. C., Gerber, J. S., Engstrom, P. M., Mueller, N. D., Brauman, K. A., Carlson, K. M., … Siebert, S. (2014). Leverage points for improving global food security and the environment. *Science*. https://doi.org/10.1126/science.1246067.

Wigan, M. R., & Clarke, R. (2013). Big data's big unintended consequences. *Computer*. https://doi.org/10.1109/MC.2013.195.

Witte, J. H. (2017). The blockchain: A gentle introduction. *SSRN Electronic Journal*. https://doi.org/10.2139/ssrn.2887567.

Wood, G. (2014). Ethereum: A secure decentralised generalised transaction ledger. *Ethereum Project Yellow Paper*. https://doi.org/10.1017/CBO9781107415324.004.

World Bank. (2016). Agriculture, value added (% of GDP). *Data*. https://doi.org/10.1179/136485908x337463.

Yahya, N. (2018). Agricultural 4.0: Its implementation toward future sustainability. In *Green energy and technology*. https://doi.org/10.1007/978-981-10-7578-0_5.

Yusuf, A. A., Peter, O., Hassanb, A. S., Tunji, A. L., Oyagbola, I. A., Mustafad, M. M., & Yusuf, D. A. (2019). Municipality solid waste management system for Mukono District. *Procedia Manufacturing, 35*, 613–622. https://doi.org/10.1016/j.promfg.2019.06.003.

Zambon, I., Cecchini, M., Egidi, G., Saporito, M. G., & Colantoni, A. (2019). Revolution 4.0: Industry vs. agriculture in a future development for SMEs. *Processes*. https://doi.org/10.3390/pr7010036.

Zeng, H., Chen, X., Xiao, X., & Zhou, Z. (2017). Institutional pressures, sustainable supply chain management, and circular economy capability: Empirical evidence from Chinese eco-industrial park firms. *Journal of Cleaner Production*. https://doi.org/10.1016/j.jclepro.2016.10.093.

CHAPTER

10 Economics and risk assessment of new technologies in agrowaste diversity

10.1 Introduction

New demands and regulations that support implementation of nascent technologies for environmental impact reduction seek the most effective approach (cost-effective, high-performance, and operational capabilities) with long-term competitive advantage. The need for a high level of understanding linked to the requirements with regard to the implementation of several sustainability considerations cannot be ignored (Dong et al., 2016; Korobar & Siljanoska, 2016). The aforementioned premise can be justified, as the negative socioenvironmental impact continues to increase due to the lack of scope to identify the challenges that are predominant over what is considered to be a threat. During the first industrial revolution (discovery of coal), the immediate climatic effect was not significant, although the operations left the natural surroundings in shamble and disarray (Onu & Mbohwa, 2018d). Improvement was therefore embraced once an understanding of the full conception of climate change became a household issue and the need for regulated operations (Horbach, Rammer, & Rennings, 2012; Khalili, Duecker, Ashton, & Chavez, 2015).

This led to the ideation of sustainable initiatives, backed by policies, and technological requirements alike. Thus, a decision was followed, to interest the global communities in the best interest of the planet to become responsible in a sustainable manner (Brundtland, 1987). The voluntary commitment and compliance to a specific agreement that quarantined sustainable benefits and climate change mitigation were sorted after. As the cost of financing the appropriate services and systems required to meet sustainability expectations falls within reconcilable benefits, the confidence to implement agrowaste management becomes high. It is easy to relate this with the fact that imminent low-cost recycling technologies are only sufficient in the short run. For example, flexible designs of robust bioplant converter are required to operate efficiently to meet the various challenges of agrowaste management. More so, the cost of reconciling the long-term effects caused by the lack of waste management far exceeds the immediate cost of avoiding the waste and its treatment for proper disposal or recycling for gainful biofuels production.

Agricultural Waste Diversity and Sustainability Issues. https://doi.org/10.1016/B978-0-323-85402-3.00004-8

Again, the production of a viable biofuel from potentially significant agricultural resources means high development costs (new challenges). The diversity in the agricultural extension for sustainable energy production, as earlier suggested, through technology advancement draws concern to its progress as to whether or not it leads to the production of viable energy resources. Consequently, the incentives required to achieve sustainability from agricultural development perspective, and techniques (green initiatives) must be available all through the farming operation, from production through processing, distribution, and retailing, up to the end of consumption with no waste disposal concern (Kumar et al., 2017; Tan, Lee, & Tan, 2019). Hence, the main reason to intensify waste management operations and schemes that will manage trash collection, product creation, and informatization of the agroindustrial sector to operate on a zero-waste basis.

The aforementioned demand can be met by combining timely access and application of technological knowledge, supported by the essential policies, framework, and financial structure if or when available to ameliorate the threats of agricultural and food industry wastes (Royal Society of London, 2009). Recently, the advancement of ICT in developing countries for sustainable business advisory services to farmers is increasing, as such will promote newly developing strategies and technology integration (Binswanger-Mkhize & Savastano, 2017; Muzari, Gatsi, & Muvhunzi, 2012; Schut et al., 2016). Moreover, the synergy of education and policies that can stimulate investments in state-of-the-art technology systems is lacking in Africa and requires great attention from a regional collaborative context. Hence, it will lead to a transformative change, a change in the understanding of resource management, production planning, and business protocols for effective agrowaste management.

A training process is, in essence, required for the restructuring of African mentality on conservatism, which should start from the "secondary schools," up to the higher, university/college, and in line with sustainable approaches to apply the emerging technological skill set to improve livelihood through agriculture diversity. There is, therefore, the need to rethink sustainable policies that will apply to Africa in the context of socialism and boundaryless approach to development, sharing technologies, human resources, and implementing intercountry regulations that appeal to agricultural development. Imaginations around smart wastes management and biodiversity in line with the Industry 4.0 trend can be encouraged to address specific challenges and limitations (Onu & Mbohwa, 2019b). As such, the corporate goal must be toward the success of Africa, for Africa, on green innovation and ecodiversity.

Viable farm resources versus agrowaste conversion for sustainable energy

The latest argument in waste management disposition contests the actions of transforming waste resources from agriculture to sustain the farm (feed animals and nitrify the soil) instead of its use in diverse fields. However, the reality before us is that livestock can only graze so much, as what is viable to them, while the remains are converted to value for money. The advancement of green technology innovation has been suggested as a critical contention to promote resource conservation (Hosenuzzaman et al., 2015; Moore, 2013). In addition, developing countries and Africa in perspective must diversify their economies through investment and the advancement of technoinnovation with a focus to improve resource conservation and agricultural extension. Mindful that the economic stability and policies of each nation will deploy the necessary strategy, as per ecoefficient, green innovation action aimed at effective agricultural development. As such, it calls for deliberations by institutions and governments, including investors/stakeholders for future sustainable development, with a focus on societal, social, economic, and environmental consideration. This will promote more awareness about the conservation of our ecosystem.

The focus of this chapter is to provide inference for the promotion of new possibilities that extend innovation capabilities, which are desired to ensure a sustainable future for resource control and agricultural development globally, and within the sub-Saharan Africa region. As such, the consideration as to the exploitation of natural resources through conservative practices, and how the proliferation of technology innovation (Sustainable Development Goals, SDG-8) affects other SDGs, forms the basis for the subsequent discussion.

Where is sub-Saharan Africa in all of this?

Among the 54 countries in Africa, the sub-Saharan region is made up of 46 countries. These countries are multiculturally diversified and comprising of some of the most economically challenged nations in Africa, while at the same time enlisted as rare mineral-rich countries. Mostly, the countries in the region have experienced civil unrests due to tussle and hustle over natural resource endowment and political extremism. Others such as South Africa, Nigeria, Kenya, and Ghana, to list a few, are performing economically and exploiting their agricultural capabilities to contribute to their country's economy. Regardless, another born-of-contention is observed. Small-scale farming appears to be one part of the sustainable development agenda, which received very little attention, whereas its activity is predominant in the poorest of regions, even within the poor rural areas of the sub-Saharan Nations, where it provides enough to sustain life for the people caught in that space and time. The situation as it may appear poses developmental implications and high risk to the immediate ecosystem and the global climate condition at large.

Agricultural operations in these rural settlements, and even in the urban areas of some developing countries, are unsustainable, not monitored, regulated, or legally informed, which has led to detrimental environmental, health, and social implications (Onu & Mbohwa, 2018c). The focus, therefore, is on the necessary interventions to salvage the imminent danger. In essence, what is the most sustainable approach to address the already existing saddening situation of poverty; health and well-being; decent work and economic growth; life on land; and peace and justice, and strong institution (SDG-1,3,8,15 and 16), in the sub-Saharan African region. Collaboration between nations of the region can promote the expansion of agricultural resources for their joint development, thus sharing/contributing knowledge for excellence and risk operations. Similarly, capacity building and infrastructural financing within the areas of agricultural wastes conversion to collectively strengthen waste management activities can be guaranteed (Ellabban, Abu-Rub, & Blaabjerg, 2014).

The derivatives from the collective interboarder projects (bioenergy from agro-waste plants) can be used to sustain the low-income countries. For example, the supply of electricity to the neighboring border, villages, and islands would be based on whether it is their prerogative to access the electric power supply. Additionally, sharing of information and technological capabilities within the region can increase the focus on coordination to track progress and inform the decision that will further lead to the setting-up of the necessary mechanism and implementation of the required protocol/policies for extensive agriculture diversity. The current proposition is imagined as a centralized approach to development, popularly perceived to be a complicated strategy to establish and maintain due to the political and cultural complexity of the different countries in the region. However, matters of this nature desperately need to be addressed, such that the potential outcome is desired by all involved, and be guided by regulatory policies.

The cleaner production and alternative energy development strategies to support sustainable development via the framework that benchmarks satiable standards in achieving agrowaste diversity vary from one place to the other depending on the law of the region. The recommendation is to have sustainable strategies that look beyond traditional practices and crude methods and pertains to effective resources-processing and distribution, which conforms to energy-efficient routines (Onu & Mbohwa, 2019; Smith, Williams, & Pearce, 2015). The growing managerial strategies to make aware consumers through informative workshops, training programs, and awareness campaigns organized to enlighten small groups have been proposed (Filimonau & Gherbin, 2018). Also, a gradual incentivization process that promotes innovation and developmental acceleration or improvement on waste reduction at farms up to the final stage of product consumption is advised. Africa can learn from the standardization, regularization, and implementation of green innovation strategies identified by notable interest groups and international bodies that are bent to promote positive socioeconomic and environmental achievement (OECD/IEA & IRENA, 2017).

10.2 Drivers and implications of technoinnovative appraisal in agrowaste management

Development practitioners, research experts, company CEOs (food process), and other stakeholders are expected to come together to evaluate and identify requirements that are necessary to strategies that will usher technical appreciation and behavioral change to waste management. Hence, the cost consideration for the actualization of these strategies cannot be overlooked (Onu & Mbohwa, 2018b). The cost of conducting project feasibility, including the cost of setting-up the financial structure to fund the project, negotiation, supervision, and technical overview, or the cost of certification, registration, compliance, and permit services, draws attention to the economics of waste management. Furthermore, concerning nature and type of agrowaste conversion technology (biodigester, pyrolysis, incinerator, etc.), the cost of operational services, including hidden and direct/book costs, within the agrowaste beneficiation budget (covers across labor cost, utility and energy expenditure, additives and consumables, communication and awareness, and insurance cost to mention but few), lowers revenue generation in agrowaste diversity (Yousuf, 2012).

Importantly, the era of the fourth industrial revolution seeks the most readily available mechanism to integrate agricultural operations to ensure sustainability in agricultural exploration (Puyol et al., 2017), which comes at a high price. However, the economics of scale for operations toward sustainable development raise the awareness, to overcome the technicalities involved to transform agrowastes into wealth. Newly developing trends in sustainable fuel production to decarbonize future combustion operations present an opportunity for the expansion of several wastes' conversion technologies (Onu & Mbohwa, 2018a). As such, more understanding concerning the socioeconomic and environmental advantages of agrowaste diversity can be prioritized, as agricultural development activities become the prerogative for implementation of advanced techniques (Ferranti, 2019; FAO, 2019). Therefore, it becomes an area of attention for governments and other major stakeholders.

Notably, as new opportunities set in, regarding businesses within the agrowaste management framework, collection and transport, processing and treatment, and utilization, before disposal stages, the development of new policies is quintessential. This will promote agricultural waste management and prevention through recycling/reuse approach. Biological pretreatment and improvement in its processes now focus on sustainable and robust technologies that are higher efficiency, cost-effective, and easily operable. As such, operational performances can be improved through optimization of the process involved, contributing to the reduction of greenhouse gas emission (Onu & Mbohwa, 2018b). Furthermore, it will offer people the opportunity to create and use, or manage raw materials and products with a high sense that considers sustainable production and consumption. Recent disposition involves thinking new pathways for conservable practices in a broad sense of applicability and acceptance of agro-based resources (food/cash crops, meat, and fish, from abattoir and food processing plants) to drive sustainable intervention and the promulgation of innovative technology (Onu & Mbohwa, 2018a; Val del Río, Campos Gómez, & Mosquera Corral, 2016).

10.3 Toward integrated waste management systems for sustainable energy sovereignty in sub-Saharan Africa

The outlook toward agricultural diversity in the near future draws interest to effective management techniques; otherwise, it may lead to catastrophic challenges of preparing landfill to accommodate the different types of wastes (based on toxicity and biodegradability), dealing with smells and the adverse effect of greenhouse gas emission, and toxic runoff, which affects useful fertile lands and water ecosystem (Gomiero, 2018; Pattanaik, Pattnaik, Saxena, & Naik, 2019; Pollet, Staffell, & Adamson, 2015). The toxicity nature of industrial wastes depends on the area or type of corporation from which the waste is generated. A very high degree of risk related to a specific kind of hazardous waste comes from sensitive mining operations and projects involving the use of radioactive material. These radioactive elements have a significant effect on humans' health and environmental stability, over a lengthy period, cutting through different generations. In most cases, they are discharged untreated in the open fields, leading to increasing causes of death/harm to the living and severe implication on the world's ecology, owing to the lack of proper sanitary landfills to manage the wastes (Demirbas, 2011).

There is good news, however, as environmental safety practitioners and experts around the world continue to search and identify high-risk project regions, municipalities, and cities with increasing waste generation tendencies to roll out the mechanism and technological imperative to curtail the awaiting menace without compromising on servitude (Song, Wang, Li, & Duan, 2012). Different innovative approaches exist: educating and implementing new technology, improving old infrastructure to feet new design requirements, and cultural indifference to gainful utilization of the agricultural wastes (International Energy Agency, 2016; United Nations (UN)., 2015). To advance this intervention, nascent contributions and impartation to the concept of agrowastes diversity are required. In the long run, feeding styles need to change, and if possible, revolutionize product design to meet eco-friendly biodegradable standards. Conveniently, by-products of agrowastes/ashes can be utilized in the development of composite materials to lead zero-wastes operation (Abdul Qayoom Tunji Lawal et al., 2019; Onu, Abolarin, & Anafi, 2017), or as raw materials for clean energy production.

The trends noticed among sub-Saharan, African nation, and other developing countries reveal the inability and inadequate capacity in terms of technical capabilities to properly manage wastes using the crude traditional methods from the treatments point-of-view to the biobeneficiation for economic and environmental advantages. The availability of biomasses, from viable plants (of the different generation of energy feedstocks) and animal wastes, is a pathway to enjoy the wastes to energy potential for production of biofuels in various applications, ranging from oils for lubrication, source of electricity generation, and production heat (Kuhad & Singh, 2013). As part of the economic importance of agrowaste management, which covers three major areas, production of clean energy, reduction of greenhouse gases, and waste reduction to landfills, the development of the necessary technologies and strategies toward sustainable energy production cannot be overemphasized and requires intensive scientific input from the context of developing Africa's green potential (from grassroot level, in the villages, to the level of advanced/organized farming).

While, the role of agricultural operations linked to wastes recycling, as per effective resource management, holds great potential for sustainable clean energy.

Consequently, the process is a contribution, and conscious attempt to ameliorate future disaster. The mapping process of sustainable development agendas in comparison with some Africa's national development plan targeting ecological infrastructures, critical for the socioeconomic growth and development, is continually being reviewed (Cumming et al., 2017; Giwa, Alabi, Yusuf, & Olukan, 2017; Government of Uganda (a), 2015). From our observation, we learn that the global drive toward the optimization of energy systems and the implementation of conservative practices to mitigate climate change and resolve anthropogenic technicalities within energy-intensive industries has shown great potential. The option for safe, reliable, and utterly green technological innovation for heat and electricity generation to the rural and urban settlement must, therefore, explore the agrowaste conversion pathway.

The aforementioned satisfies the environmental, technological, economic, social, and cultural sustainability euphoria to support sustainable development goals. Moreover, agroindustrial operations have the potential to cause unparalleled development through clean energy production and contribute to human health and livelihood through inclusive strategies (Dick, 2016; Ng, Xu, Yang, Lu, & Li, 2018). Innovation in fields of biodegradable waste exploration can lead to competitiveness and result in excellence for different operations having low-risk tendencies and a positive impact on sustainability improvement (Desai, Dalvi, Jadhav, & Baphna, 2018). More so, effective mechanisms and procedures that will ensure safe transport and delivery from the agricultural perspective, to where the agrarian produces are needed, will complement the campaign. Governmental grants, or the tendency to allow for free, the recovery and utilization of degraded land areas, and compensation for the restoration of damaged lands for agricultural purposes will contribute to quality food production, which is cheap and readily available.

10.4 Conclusion and perspectives

This chapter examines the interlink between sustainability and the ideals for economic progress amid the risk and uncertainty that threatens the advancement of agricultural development in Africa. It focuses on the approaches with regard to technological proliferation, regulatory policies, and scientific perspectives (education and research), which played a pivotal role in waste management and agricultural development. The authors are optimistic that the problems due to agricultural development can be resolved in developing countries, if equal opportunities to access loans and funds, or when the bureaucratic bottlenecks involved in accessing government incentives are justified (No favoritism). As such, ease to which information is obtained by the wealthy, poor, political elite, or local farmers (equality), concerning agriculture diversity must reflect transparency and diligence to the rule of law in the respective region/state/nation. The educational aspect (information and communication) of agricultural development requires organizing seminars, workshops, and conducting a comprehensive study, among other strategies. In essence, teaching and learning can support the framework for sustainability advancement. Additionally, overcoming the "stereotype syndrome" linked to female education and participation in activities that promises productivity is timely in Africa since all potential efforts are needed to achieve the desired success.

Also, cultural misbelief concerning seasonal considerations, forbidden resources exploitation, or application of what type of strategy/technology to be used for what purpose must be derided. Socioeconomic empowerment is a factor that corroborates the number 8th sustainable development goal and, thus, emphasizes that a high level of productivity can be the achievement through technological innovation, links the first, second third, fourth, and fifth agenda. As such, higher levels of productivity will ensure well-being, related to the eradication of poverty, which takes away hunger, and facilitates healthiness, offering equal opportunities to forester education, irrespective of gender, whereas the development of sustainable technologies has suffered serious proliferation challenges (Africa) due to misconception, ignorance, and fear of anticipated side effects, or return on investments, as stakeholders perceive it, disregarding the environmental impact, causative by poor waste management, over a long period. Sustainable interventions must focus on specific solutions to environmental challenges and be backed by regulations that support the development and propagation of new technologies, compliance to set standards that subscribe to professional practices.

The context of education in the 6th, 7th, and 15th agenda is to raise awareness to promote sanitation and resource/water management, efficient energy usage, and prevention of variation on the ecosystem. Inter-Africa partnership in all front concerning economic progress amid the risk and uncertainty that threatens the advancement of agricultural development is important because it will bridge the gap on income inequality and promote economic convergence, leading to a prosperous Africa and sub-Saharan Africa in perspective. In closing, this chapter has provided insight into the relevance of propagating advanced technology initiatives as conceptualized by notable researchers to exploit the opportunities of sustainable interventions and agricultural waste diversity. Coverage of the sub-Saharan context is also reviewed. It is worthwhile for prospective studies to consider the qualitative impact of the objectives of this study in different subregions: West Africa, East Africa, and Southern Africa separately. Hence, this chapter determines the success propensity of nascent technology and sustainable initiative for future agriculture wastes diversity and sustainability improvement.

References

Abdul Qayoom Tunji, Lawal, Ninsiima, E., Odebiyib, O. S., Hassan, A. S., Oyagbola, I. A., Onu, P., & Danjuma., A. (2019). Effect of unburnt rice husk on the properties of concrete. *Procedia Manufacturing, 35*, 635–640. https://doi.org/10.1016/j.promfg.2019.06.006.

Binswanger-Mkhize, H. P., & Savastano, S. (2017). Agricultural intensification: The status in six African countries. *Food Policy, 67*, 26–40. https://doi.org/10.1016/j.foodpol.2016.09.021.

Brundtland, G. H. (1987). Our common future (brundtland report). In V. Hauff (Ed.), *United Nations commission*. Oxford University Press.

Cumming, T. L., Shackleton, R. T., Förster, J., Dini, J., Khan, A., Gumula, M., & Kubiszewski, I. (2017). Achieving the national development agenda and the sustainable development goals (SDGs) through investment in ecological infrastructure: A case study of South Africa. *Ecosystem Services, 27*, 253–260.

Demirbas, A. (2011). Waste management, waste resource facilities and waste conversion processes. *Energy Conversion and Management, 52*, 1280–1287.

Desai, Y., Dalvi, A., Jadhav, P., & Baphna, A. (2018). Waste segregation using machine learning. *International Journal for Research in Applied Science and Engineering Technology.*

Dick, E. (2016). *Urban governance for sustainable global development: From the SDGs to the new urban agenda.* German Development Institute Briefing Paper.

Dong, L., Fujita, T., Dai, M., Geng, Y., Ren, J., Fujii, M., … Ohnishi, S. (2016). Towards preventative eco-industrial development: An industrial and urban symbiosis case in one typical industrial city in China. *Journal of Cleaner Production.* https://doi.org/10.1016/j.jclepro.2015.05.015.

Ellabban, O., Abu-Rub, H., & Blaabjerg, F. (2014). Renewable energy resources: Current status, future prospects and their enabling technology. *Renewable and Sustainable Energy Reviews, 39*, 748–764. https://doi.org/10.1016/j.rser.2014.07.113.

FAO. (2019). *Save food: Global initiative on food loss and waste reduction.* Food and Agriculture Organization of the United Nations.

Ferranti, P. (2019). The united nations sustainable development goals. In J. R. Anderson, et al. (Eds.), *Encyclopedia of food security and sustainability.* Encyclopedia.

Filimonau, V., & Gherbin, A. (2018). An exploratory study of food waste management practices in the UK grocery retail sector. *Journal of Cleaner Production, 167*, 1184–1194. https://doi.org/10.1016/j.jclepro.2017.07.229.

Giwa, A., Alabi, A., Yusuf, A., & Olukan, T. (2017). A comprehensive review on biomass and solar energy for sustainable energy generation in Nigeria. *Renewable and Sustainable Energy Reviews, 69*, 620–641. https://doi.org/10.1016/j.rser.2016.11.160.

Gomiero, T. (2018). Agriculture and degrowth: State of the art and assessment of organic and biotech-based agriculture from a degrowth perspective. *Journal of Cleaner Production, 197*, 1823–1839. https://doi.org/10.1016/j.jclepro.2017.03.237.

Government of Uganda (a). (2015). *Second national development plan - Uganda.* National Planning Authority Uganda.

Horbach, J., Rammer, C., & Rennings, K. (2012). Determinants of eco-innovations by type of environmental impact - the role of regulatory push/pull, technology push and market pull. *Ecological Economics, 78*, 112–122. https://doi.org/10.1016/j.ecolecon.2012.04.005.

Hosenuzzaman, M., Rahim, N. A., Selvaraj, J., Hasanuzzaman, M., Malek, A. B. M. A., & Nahar, A. (2015). Global prospects, progress, policies, and environmental impact of solar photovoltaic power generation. *Renewable and Sustainable Energy Reviews, 41*, 284–297. https://doi.org/10.1016/j.rser.2014.08.046.

International Energy Agency. (2016). *World energy outlook 2016.* International Energy Agency. http://www.iea.org/publications/freepublications/publication/WEB_WorldEnergyOutlook2015ExecutiveSummaryEnglishFinal.pdf.

Khalili, N. R., Duecker, S., Ashton, W., & Chavez, F. (2015). From cleaner production to sustainable development: The role of academia. *Journal of Cleaner Production, 96*, 30–43. https://doi.org/10.1016/j.jclepro.2014.01.099

Korobar, V. P., & Siljanoska, J. (2016). Challenges of teaching sustainable urbanism. *Energy and Buildings, 115*, 121–130. https://doi.org/10.1016/j.enbuild.2015.04.049.

Kuhad, R. C., & Singh, A. (2013). Biotechnology for environmental management and resource recovery. In R. C. Kuhad, et al. (Eds.), *Biotechnology for environmental management and resource recovery.* Springer.

Kumar, S., Smith, S. R., Fowler, G., Velis, C., Kumar, S. J., Arya, S., ... Cheeseman, C. (2017). Challenges and opportunities associated with waste management in India. *Royal Society Open Science*. https://doi.org/10.1098/rsos.160764.

Moore, D. R. (2013). Technologies. In *Green energy and technology*. Spinger.

Muzari, W., Gatsi, W., & Muvhunzi, S. (2012). The impacts of technology adoption on smallholder agricultural productivity in sub-saharan Africa: A review. *Journal of Sustainable Development, 5*(8). https://doi.org/10.5539/jsd.v5n8p69.

Ng, S. T., Xu, F. J., Yang, Y., Lu, M., & Li, J. (2018). Necessities and challenges to strengthen the regional infrastructure resilience within city clusters. *Procedia Engineering, 212*, 198–205. https://doi.org/10.1016/j.proeng.2018.01.026.

OECD/IEA, & IRENA. (2017). Perspectives for the energy transition: Investment needs for a low-carbon energy system. In *International energy agency*.

Onu, P., & Mbohwa, C. (2018a). Future energy systems and sustainable emission control: Africa in perspective. *Proceedings of the International Conference on Industrial Engineering and Operations Management*, 793–800.

Onu, P., & Mbohwa, C. (2018b). Green supply chain management and sustainable industrial Practices: Bridging the gap. *Proceedings of the International Conference on Industrial Engineering and Operations Management*, 786–792 (Washington DC).

Onu, P., & Mbohwa, C. (2018c). Sustainable oil exploitation versus renewable energy Initiatives: A review of the case of Uganda. *Proceedings of the International Conference on Industrial Engineering and Operations Management*, 1008–1015 (Washington DC).

Onu, P., & Mbohwa, C. (2018d). Correlation between future energy systems and industrial revolutions. *Proceedings of the International Conference on Industrial Engineering and Operations Management*, 1953–1961 (Pretoria/Johannesburg).

Onu, P., & Mbohwa, C. (2019a). A sustainable industrial development approach: Enterprise Risk Management in view. *IOP Conference Series: Materials Science and Engineering*, Article 022094.

Onu, P., & Mbohwa, C. (2019b). Industrial energy conservation initiative and prospect for sustainable manufacturing. *Procedia Manufacturing, 35*, 546–551. https://doi.org/10.1016/j.promfg.2019.05.077.

Onu, P., Abolarin, M. S., & Anafi, F. O. (2017). Assessment of effect of rice husk ash on burnt properties of badeggi clay. *International Journal of Advanced Research, 5*(5), 240–247. https://doi.org/10.21474/IJAR01/4103.

Pattanaik, L., Pattnaik, F., Saxena, D. K., & Naik, S. N. (2019). Biofuels from agricultural wastes. In *Second and third generation of feedstocks*. https://doi.org/10.1016/b978-0-12-815162-4.00005-7.

Pollet, B. G., Staffell, I., & Adamson, K.-A. (2015). Current energy landscape in the Republic of South Africa. *International Journal of Hydrogen Energy, 40*(46), 16685–16701. https://doi.org/10.1016/j.ijhydene.2015.09.141.

Puyol, D., Batstone, D. J., Hülsen, T., Astals, S., Peces, M., & Krömer, J. O. (2017). Resource recovery from wastewater by biological technologies: Opportunities, challenges, and prospects. *Frontiers in Microbiology*. https://doi.org/10.3389/fmicb.2016.02106.

Royal Society of London. (2009). Reaping the benefits: Science and the sustainable intensification of global agriculture. In R. Lord, et al. (Eds.), *Chemical engineer*. The Royal Society.

Schut, M., van Asten, P., Okafor, C., Hicintuka, C., Mapatano, S., Nabahungu, N. L., ... Vanlauwe, B. (2016). Sustainable intensification of agricultural systems in the Central

African Highlands: The need for institutional innovation. *Agricultural Systems*. https://doi.org/10.1016/j.agsy.2016.03.005.

Smith, L. G., Williams, A. G., & Pearce, B. D. (2015). The energy efficiency of organic agriculture: A review. *Renewable Agriculture and Food Systems, 30*(3), 280–301. https://doi.org/10.1017/S1742170513000471.

Song, Q., Wang, Z., Li, J., & Duan, H. (2012). Sustainability evaluation of an e-waste treatment enterprise based on emergy analysis in China. *Ecological Engineering, 42*, 223–231. https://doi.org/10.1016/j.ecoleng.2012.02.016.

Tan, Y. S., Lee, T. J., & Tan, K. (2019). Integrated solid waste management. In *Clean, green and blue*. https://doi.org/10.1355/9789812308627-007.

United Nations (UN). (2015). *Sustainable development goals*. Knowledge Platform.

Val del Río, Á., Campos Gómez, J. L., & Mosquera Corral, A. (2016). Technologies for the treatment and recovery of nutrients from industrial wastewater. In Á Val del Río, et al. (Eds.), *Technologies for the treatment and recovery of nutrients from industrial wastewater*. IGI.

Yousuf, A. (2012). Biodiesel from lignocellulosic biomass - prospects and challenges. *Waste Management, 32*(11), 2061–2067. https://doi.org/10.1016/j.wasman.2012.03.008.

Index

'*Note*: Page numbers followed by "f" indicate figures and "t" indicate tables.'

T

U

V

W

Z

Printed in the United States
By Bookmasters